IEE CONTROL ENGINEERING SERIES 33

Series Editors: Prof. D. P. Atherton
Dr. K. Warwick

TEMPERATURE MEASUREMENT & CONTROL

Other volumes in this series

Volume 1	**Multivariable control theory** J. M. Layton
Volume 2	**Elevator traffic analysis, design and control** G. C. Barney and S. M. Dos Santos
Volume 3	**Transducers in digital systems** G. A. Woolvet
Volume 4	**Supervisory remote control systems** R. E. Young
Volume 5	**Structure of interconnected systems** H. Nicholson
Volume 6	**Power system control** M. J. H. Sterling
Volume 7	**Feedback and multivariable systems** D. H. Owens
Volume 8	**A history of control engineering, 1800-1930** S. Bennett
Volume 9	**Modern approaches to control system design** N. Munro (Editor)
Volume 10	**Control of time delay systems** J. E. Marshall
Volume 11	**Biological systems, modelling and control** D. A. Linkens
Volume 12	**Modelling of dynamical systems—1** H. Nicholson (Editor)
Volume 13	**Modelling of dynamical systems—2** H. Nicholson (Editor)
Volume 14	**Optimal relay and saturating control system synthesis** E. P. Ryan
Volume 15	**Self-tuning and adaptive control: theory and application** C. J. Harris and S. A. Billings (Editors)
Volume 16	**Systems modelling and optimisation** P. Nash
Volume 17	**Control in hazardous environments** R. E. Young
Volume 18	**Applied control theory** J. R. Leigh
Volume 19	**Stepping motors: a guide to modern theory and practice** P. P. Acarnley
Volume 20	**Design of modern control systems** D. J. Bell, P. A. Cook and N. Munro (Editors)
Volume 21	**Computer control of industrial processes** S. Bennett and D. A. Linkens (Editors)
Volume 22	**Digital signal processing** N. B. Jones (Editor)
Volume 23	**Robotic technology** A. Pugh (Editor)
Volume 24	**Real-time computer control** S. Bennett and D. A. Linkens (Editors)
Volume 25	**Nonlinear system design** S. A. Billings, J. O. Gray and D. H. Owens (Editors)
Volume 26	**Measurement and instrumentation for control** M. G. Mylroi and G. Calvert (Editors)
Volume 27	**Process dynamics estimation and control** A. Johnson
Volume 28	**Robots and automated manufacture** J. Billingsley (Editor)
Volume 29	**Industrial digital control systems** K. Warwick and D. Rees (Editors)
Volume 30	**Electromagnetic suspension—dynamics and control** P. K. Sinha
Volume 31	**Modelling and control of fermentation processes** J. R. Leigh (Editor)
Volume 32	**Multivariable control for industrial applications** J. O'Reilly (Editor)

TEMPERATURE MEASUREMENT & CONTROL

J R Leigh

Peter Peregrinus Ltd on behalf of the Institution of Electrical Engineers

Published by: Peter Peregrinus Ltd., London, United Kingdom

British Library Cataloguing in Publication Data.

Leigh, J. R. (James Ronald)
Temperature measurement and control.—
(IEE control engineering series; 33)
1. Thermometers and thermometry
I. Title II. Series
536'.5'028 QC271

ISBN 0 86341 111 8

Printed in England by Short Run Press Ltd., Exeter

Contents

Preface

Need for temperature control

Most books and academic courses on control engineering concentrate either mostly or exclusively on position- and speed-control problems. Much of what such approaches have to offer is, of course, universally applicable to any control application. However, temperature control has some special features:

(i) An asymmetry caused by the usually differing mechanisms for heating and cooling
(ii) (Almost always) The inclusion of complex nonlinear heat-transfer effects into the necessary process models
(iii) Highly application-dependent measurement problems.

Industrial process needs for temperature control

Many industries have extensive requirements for temperature control. Some examples from particular industries are given below:

Metal industries: Extraction of metals from ores, refining, casting, shaping and heat treatment all take place at high temperatures. Glass, cement, brick and pottery manufacture involve high-temperature processing.

Chemical industries: Distillation columns and reactors are two examples of processes where temperature monitoring and control is critical.

Steam production: Most factories have a steam requirement that is met by a system of boilers. Electricity generation, except at very low power levels, always involves the use of steam.

Food: Many foods require to be cooked, sterilised or stored at specified temperatures.

Biotechnology: The controlled growth of organisms in fermentation processes always requires maintaining a specified temperature to promote optimum growth.

Control of the environment

The temperature on earth varies over the approximate range −60 to +60°C and although man can survive extremes of temperature, to live comfortably and work efficiently he needs an environmental temperature of about 20°C. The requirement therefore exists to control the temperature inside a wide variety of buildings and vehicles in order to counteract the effect of wide-ranging cyclic (diurnal and seasonal) and random external temperature deviations.

Needs for minimising fuel consumption

All managements are aware of fuel costs and are particularly receptive of temperature-control schemes that offer a combination of improved process performance together with a significant reduction in fuel consumption.

The intention of the book is to treat the theory and practice of temperature measurement and control, and important related topics such as energy management and air pollution, at a level suitable for engineering and science students and in a manner designed to make the book valuable to practising engineers.

The philosophy of the book, as appropriate in an engineering text, is a compromise between fundamentals and practical guidelines. It is the author's firm belief that it is highly desirable to obtain a good insight into theoretical fundamentals (deeper than can be justified on grounds of immediate utility) before embarking on practical applications. The aim has been to produce a practically oriented text within a firm theoretical outline.

Acknowledgments

In producing this book, I have drawn on many sources, in particular on the following books (full references given within the text).

QUINN: on temperature fundamentals
GLASSMAN: on combustion fundamentals
SHINSKEY: on energy management
ZEMANKSY: on thermal principles

Many manufacturers willingly supplied details of their products. In this connection, I should like to thank:

ASEA
Calex
Geestra
Hartmann & Braun
Hewlett–Packard
Imo Precision Controls
Ircon
Jumo
Land Pyrometers
Nulect
Omega
Rototherm
Turnbull Controls

Some of the information that I have been able to use has been published previously in *Process engineering* (A Morgan–Grampian publication, UK). I am grateful to its editor, Mr. Clive Tayler, for permission to make use of the material here.

Finally, I thank my colleague, Dr. Chris Marquand, for checking the Appendix on thermodynamic fundamentals.

Short list of principal abbreviations

C_p = specific heat capacity at constant pressure
C_v = specific heat capacity at constant volume
H = enthalpy
s = complex variable associated with Laplace transformation
S = entropy
U = internal energy
W = work
Q = quantity of heat
P = pressure
V = volume
T = absolute temperature
(T_{suffix} is also used to indicate a time constant. The context makes clear what is meant).
θ = temperature
ω = frequency, rad/s
x_d = desired value (indicated by the suffix) of any variable x

Chapter 1

Definitions of temperature and heat

1.1 Initial definitions

It would be satisfying to derive a definition for temperature from first principles instead of simply insisting that a property called temperature exists. However, such a derivation is only possible provided that other basic concepts and quantities, such as heat, thermal contact and thermal equilibrium are first defined. On balance, therefore, it seems best to insist that there exists a property called temperature that is a measure of the intensity (as opposed to the extensity) of heat in a substance. It is defined as one of the state variables that describe the thermal state of a system.

Now define a physical *system* as any arbitrary set of materials, objects or devices within a clearly defined closed boundary in 3-dimensional space. The region outside the system boundary is known as the *environment* of the system. Assume that initially a system is at a different temperature from its environment, then, in general, there will be interaction between the system and its environment as a result of which energy will be transferred through the boundary, and the temperature difference between system and environment will tend to zero. The energy in transit through the boundary is defined as *heat*; i.e. heat is energy in transit, driven entirely by temperature difference. The system is said to be in *thermal contact* with its environment, and when the steady state of zero temperature difference has been reached, the state will be called *thermal equilibrium*.

A boundary that allows thermal contact between a system and its environment is called a *diathermic boundary*, whereas a boundary that completely prevents such thermal contact is called an *adiabatic boundary*. (Notice that both types of boundary completely prevent material transfer.)

1.2 Temperature measurement

From the discussion above, it is clear that systems that are in thermal equilibrium with each other have the same temperature. *Temperature measurement* involves three systems: a primary standard system, a practical measuring system and a system whose temperature is to be measured. The zeroth law of thermodynamics confirms intuition and puts measurement on a rigorous basis with its statement: 'If two systems are in thermal equilibrium with a third system then they are in thermal equilibrium with each other'.

1.3 Microscopic, thermodynamic and practical concepts of temperature

From a microscopic viewpoint temperature is a measure of the kinetic energy associated with the vibration of movement of molecules of matter. From a macroscopic viewpoint, temperature can be defined in thermodynamic terms using the concept of an idealised engine. Neither viewpoint comes near to forming a basis for defining a temperature scale against which practical measurements may be made. Consequently, there is a need for a third, practical, viewpoint. Such a practical viewpoint relates temperature to a series of well defined physical phenomena, such as the melting of pure metals. These topics are selectively developed in the following Sections.

1.3.1 Thermodynamic scale of temperature

It is a direct implication of the second law of thermodynamics that any two reversible heat engines that operate cyclically between a high-temperature source at temperature T_1 and a lower temperature T_2 will always have identical efficiencies equal to

$$1 - T_1/T_2 \left(= \frac{\text{useful energy obtained}}{\text{total energy input}} \right)$$

The implication is that, since the efficiency of an idealised heat engine depends only on the ratio of two temperatures, then, conversely, the efficiency of such an engine may be used to define the ratio between any two unknown temperatures. This concept leads to the *thermodynamic temperature scale*. This scale is valuable not in a practical sense, since it is doubtful if actual temperatures could ever be measured using an (idealised) engine in the way described. Rather, its value lies in the fact that it provides the most fundamental and theoretically sound temperature scale to underpin the miscellany of expansion thermometers and resistance, thermoelectric and radiation devices that have to be used in practice to measure temperature

1.4 Isotherms and temperature measurement

Consider a system whose physical state at any instant may be characterised by an n-dimensional vector $X_1, \ldots, X_n$. For concreteness, consider a fixed mass of gas at pressure X_1, having volume X_2 and at a temperature X_3 (the X notation is used to avoid exciting preconceived ideas concerning the gas laws). We ask the interesting question: for a fixed value of temperature X_3, what can we say about the values of X_1 and X_2? By definition, for a fixed value of X_3, the values of X_1, X_2 that are possible are restricted to a limited set that lie on an *isotherm* in (X_1, X_2) space.

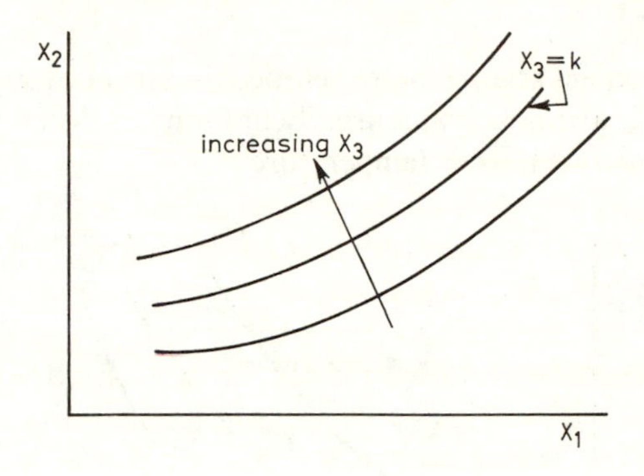

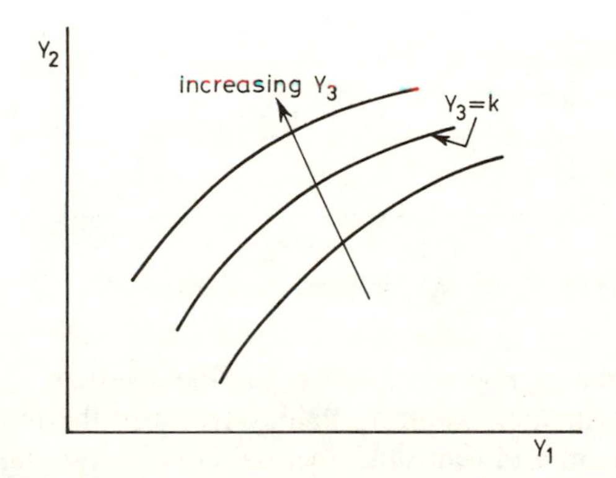

Fig. 1.1 *Two different physical systems with state descriptors (X_1, X_2, X_3), (Y_1, Y_2, Y_3), where X_3, Y_3 denote temperature*
When $X_3 = k$ and $Y_3 = k$, the two systems are in thermal equilibrium

To connect this idea with what has gone before we consider, with the aid of Fig. 1.1, two quite different physical systems characterised by state descriptors (X_1, X_2, X_3), (Y_1, Y_2, Y_3), respectively, where X_3 and Y_3 both denote temperatures. Let the two systems be in any states (X_1, X_2, K), (Y_1, Y_2, K) respectively; then they are in thermal equilibrium with each other.

To set up a temperature-measurement system we arbitrarily select one of the systems, say the system with state descriptor (X_1, X_2, X_3), as the standard and choose one of the two co-ordinates X_1, X_2 as the *thermometric variable*. Suppose, for illustration, that we fix X_1 at a convenient constant value, to produce the construction shown in Fig. 1.2. The temperature to be allocated to each isotherm is then fixed by choosing a function that relates the variable X_2 with temperature. Let X_2' be the value of X_2 when our chosen system is in thermal equilibrium with a system of ice, water and water vapour. Then X_2' corresponds to the so-called triple point of water and we could define a temperature measurement

$$\theta = \alpha(X_2 - X_2') \tag{1.1}$$

where α is a convenient scaling factor. Notice that by choosing a linear function in eqn. 1.1, there is an implicit and usually unfounded assumption that property X_2 is itself a linear function of temperature.

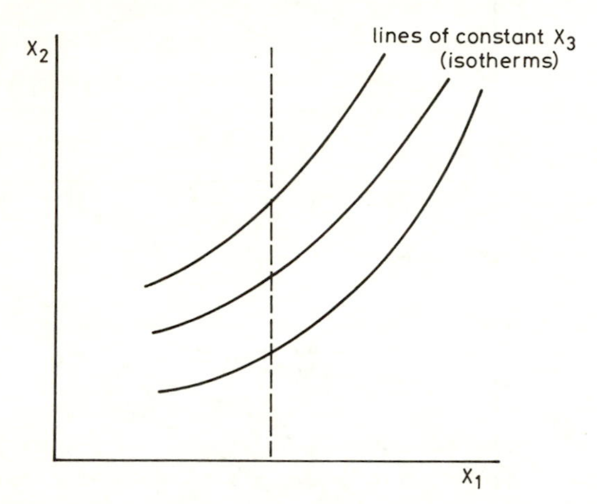

Fig. 1.2 *Illustration of the use of X_2 as a thermometric variable*

To measure the temperature of any other system, say the one with (Y_1, Y_2, Y_3) co-ordinates, we bring that system into thermal contact with our measurement system and wait until thermal equilibrium has been established. We shall then say that our measurement of the temperature Y_3 is given by $\theta = \alpha(X_2 - X_2')$ from eqn. 1.1.

The imperfections of such a system are readily apparent. Consider taking the volume of a constant mass of mercury as the thermometric variable to form the familiar mercury-in-glass thermometer. Such a device can operate over only a small temperature range compared with the range from near absolute zero up to the melting points of refractory materials that needs to be covered. Even within the restricted range of its operation, the mercury-in-glass thermometer is found to disagree with similar thermometers containing different liquids. The disagreement occurs because expansion is not necessarily a linear function. The effect is illustrated most markedly by a thermometer containing water. Since water contracts with rising temperature over the interval 0–4°C, some considerable anomalies of behaviour may be expected. These observations emphasise the need for an absolute temperature scale against which to measure the linearity of various practical temperature sensors. The gas-filled thermometer comes near to providing the absolute scale that is required.

1.5 Gas-filled thermometers

Suppose that a container is filled with a gas and then sealed, with provision made to measure the internal pressure p. Let the pressure when the container is in an ice-bath be p_0 and when in a steam bath, p_{100}. Then, experiment shows that provided the gas in the container is sufficiently rarified, $p_{100}/p_0 = 1{\cdot}36609$ *for every gas*. Thus the reading of a gas thermometer is independent of the particular gas used, and a gas thermometer therefore represents a more fundamental standardising device than does a liquid-filled thermometer. If we now define temperature T as measured by the gas thermometer by the relation

$$\frac{T}{T_0} = \frac{p}{p_0}$$

and

$$T_{100} - T_0 = 100$$

where p is the measured pressure at the unknown temperature T and the second equation is simply a statement that the interval between steam point and triple point shall be divided into 100 degrees, then we obtain the result that

$$T = 273{\cdot}16 \frac{p}{p_0}$$

The gas thermometer provides an absolute scale because it provides a true zero. However, because practical gas thermometry is difficult to apply, *fixed temperature points* have been defined and agreed internationally in terms of the freezing points of pure substances. Practical temperature sensors may then be calibrated in terms of those fixed points, although there is still the problem of nonlinear behaviour between fixed points.

1.6 Practical temperature scale defined in terms of fixed points

Instruments that can measure absolute temperatures exist only in primary calibration laboratories. Accordingly, fixed points that can be reproduced accurately have been agreed internationally to have numerical values that agree closely with the absolute temperature scale. Six primary fixed points were agreed by an international conference in 1968. They are:

	Temperature degs C
Boiling point of oxygen	−182·962
Triple point of water (i.e. where water, ice and water vapour co-exist).	0·01
Steam point	100
Freezing point of zinc	419·58
Freezing point of silver	961·93
Freezing point of gold	1064·43

A number of secondary fixed points have also been defined. Interpolation between the fixed points is performed in a prescribed manner as follows.

For temperatures up to 630·74°C, the practical scale is defined in terms of measurements from a platinum-resistance thermometer in conjunction with a set of given interpolating functions that together span the temperature range.

For temperatures from 630·74°C to 1064·43°C, the scale is defined in terms of measurement from a platinum/90% platinum, 10% rhodium thermocouple, together with a simple parabolic interpolating function.

To illustrate how the interpolation proceeds, we let the EMF of the thermocouple e be given by

$$e = a + b\theta + c\theta^2$$

The contents a, b, c are determined from tests at the silver, gold and antimony fixed points. Any intermediate temperature is then determined from e using the equation.

For temperatures above 1064·43°C, an optical pyrometer is used in conjunction with Wien's law of black-body radiation.

If a set of different types of measuring device has to be chosen as fundamental standards for different temperature ranges, there will clearly be problems in smoothly connecting the separate scales to form a well constructed overall practical scale. This problem of defining a practical temperature scale based on several different measuring systems has exercised international standards committees for many years. In particular, it should be noted that some of the interpolation techniques agreed by international standardising committees are quite complex. Quinn (1983) should be consulted for details.

1.7 Summary

To summarise, the underlying absolute thermodynamic temperature scale measured in degrees Kelvin is approximated for practical purposes by a degree Celsius scale that is defined with the aid of fixed points, interpolating instruments and interpolating functions. It is interesting to note that the relation

$$°C = °K - 273{\cdot}15$$

is only approximately valid, with highest accuracy in general at the primary fixed points. (This point is only of academic interest, since the errors involved are extremely small, but an appreciation that the equation is an approximation, however good, helps an understanding of the fundamentals.)

1.8 Further reading

QUINN, T. (1983): 'Temperature' (Academic press).

A further discussion of fundamentals with a list of supporting literature is to be found in the Appendix.

Chapter 2

Brief review of simple indicating thermometers, for low-cost temperature measurement in the −30°C to +400°C range

2.1 Thermometers

A number of types of thermometer is available for temperature indication as follows:

Liquid-in-glass thermometers: The most familiar type: they can be used over the range −30°C to 200°C.

Bimetallic thermometers: These devices have a robust cylindrical stem, which is immersed in the fluid to be monitored, and a circular dial indicating the temperature. Models are available to cover the range −30°C to 400°C. The

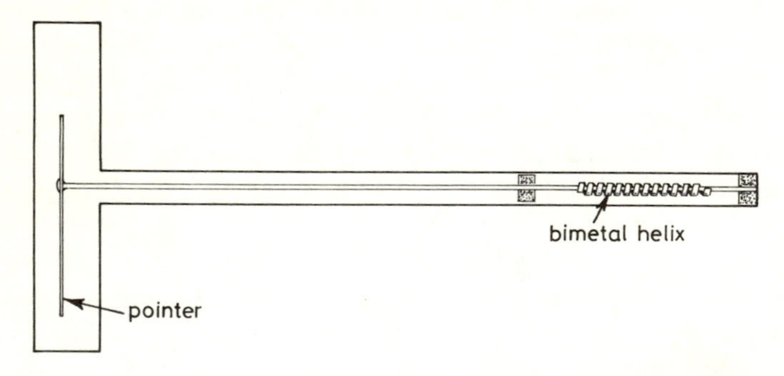

Fig. 2.1 *Principle of bimetallic dial thermometer*

principle is shown in Fig. 2.1 and the external appearance of a typical device in Fig. 2.2.

Mercury-in-steel thermometers: The expansion of mercury in a sealed steel bulb is transmitted along a narrow-bore tube to a Bourdon tube that drives a temperature indicator. Available instruments together span the range −30°C to 400°C. The principle of operation of a typical device is shown in Fig. 2.3.

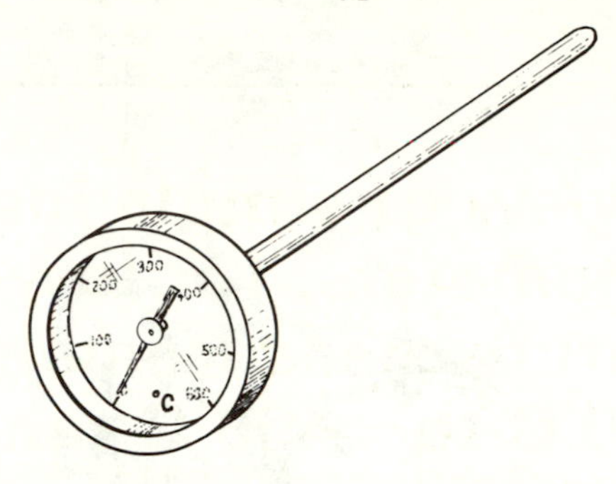

Fig. 2.2 *Bimetallic dial thermometer*

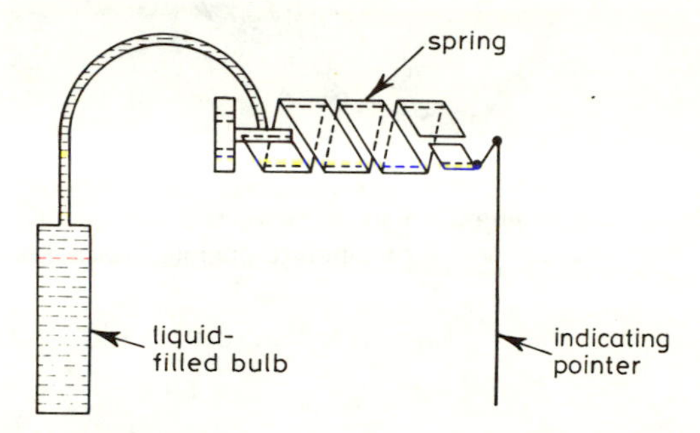

Fig. 2.3 *Principle of the Bourdon tube*

Vapour-pressure thermometer: A vapour-pressure thermometer features a bulb that is immersed in the fluid to be measured. The bulb is joined by a flexible tube to an indicating dial. The principle is that a liquid in equilibrium with its own vapour exerts a pressure that increases with temperature and that drives the remote indicator. Available instruments together span the temperature range −10°C to 200°C. The appearance of a commercial vapour-pressure thermometer is shown in Fig. 2.4.

Most of the above types of thermometer may be obtained in versions that operate contacts or that provide an electrical signal. Thus, all devices are potentially able to operate in automatic control loops.

2.2 Temperature indicating crystals, paints, labels etc.

A wide range of simple temperature-sensitive indicating materials are available for use in practical experiments or for routine quality control. Some of the materials available are:

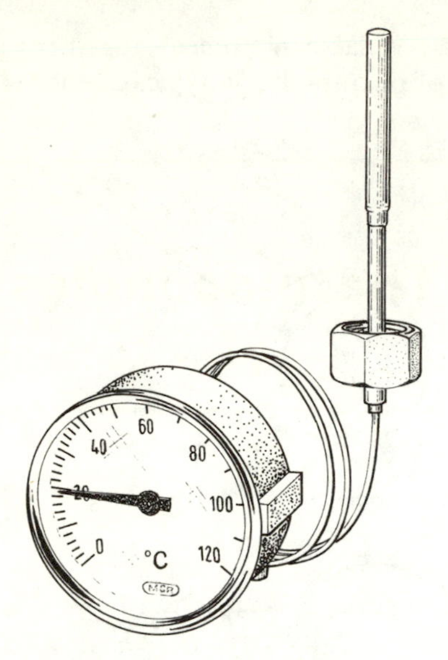

Fig. 2.4 *Vapour-pressure thermometer*
The bulb may be mounted up to 35 m from the measuring instrument. Vapour pressure thermometers may be obtained to operate over the range −10°C to 200°C

Pellets and crayons that melt at known temperatures (40°–1400°C range available)

Labels that change temperature irreversibly once a particular temperature has been exceeded. These are useful for ensuring that electrical or mechanical devices have not overheated. A single label may incorporate a range of temperature-sensitive elements so that the highest temperature reached to date may be read off at any time. Labels that change reversibly are also available.

Numeric liquid crystals that indicate the temperature in digits along a linear scale. These devices can be obtained in the form of a thin piece of adhesive-backed plastic. Liquid crystals operate only over a narrow range 0–60°C.

2.3 Comment

Chapter 3 contains sections on the speed of response of thermocouples. Many of the conclusions are common for any temperature-measurement probe and they can usefully be read with thermometers in mind.

Chapter 3

Thermocouples

3.1 Thermoelectricity

Suppose that two wires composed of different materials A and B are twisted together at their ends to form a closed loop as in Fig. 3.1. Then if the two junctions are at different temperatures θ_1, θ_2, it is found that a thermally generated EMF, dependent only on the compositions of the two wires and on $\theta_2 - \theta_1$, is acting in the loop. Provided that the wires are of accurately known composition, the generated EMF may be used as a measure of temperature difference between the twisted junctions.

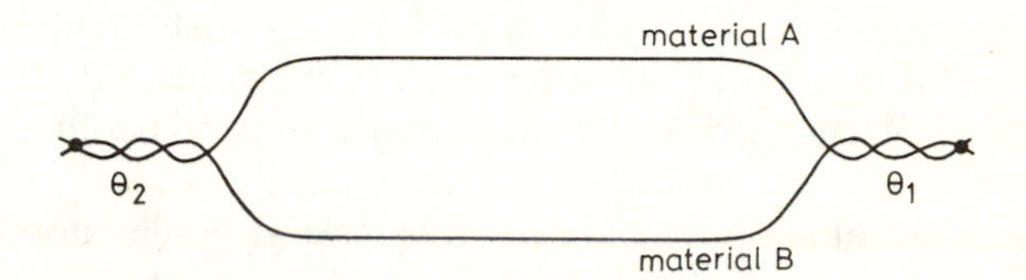

Fig. 3.1 *Thermocouple circuit*

3.2 Theoretical foundations

Physically, thermoelectric EMFs arise because, where there is a temperature gradient in a material, electrons experience a force that is directed down the temperature gradient. Where the material is a conductor, the electrons are free to move to an extent dependent on the electrical conductivity, and consequently a thermally generated potential difference will appear between the ends of any conductor that is subject to a temperature gradient. Consider such a conductor (Fig. 3.2) to have reached thermal and electrical equilibrium. More electrons are present at the cold end than at the hot end because of the force referred to earlier. Equilibrium is reached when this thermal force down the temperature gradient is balanced by the repulsion force from the charge of electrons accumulated at the cold end. Consider any point such as X in the Figure. The number of electrons crossing a line through X in any time interval must sum to zero at equilibrium.

However, those electrons crossing in the direction hot to cold have higher velocity, and hence higher energy, than their counterparts moving in the opposite direction, and by this mechanism heat conduction down the temperature gradient is achieved.

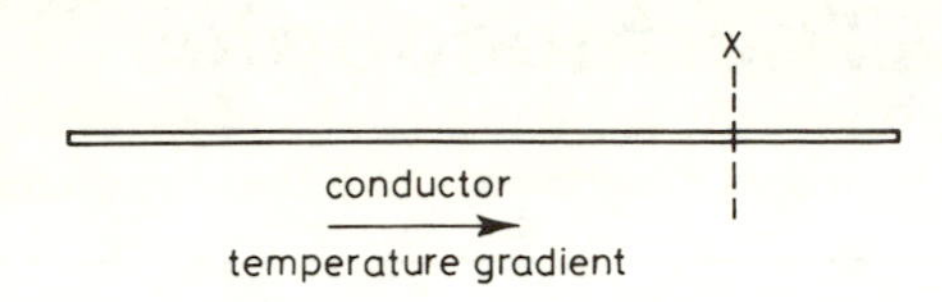

Fig. 3.2 *Conductor exposed to a temperature gradient*

When we come to measure the thermoelectric EMF that we have just described a difficulty arises. In order to measure the EMF, we must form a closed electrical loop. Suppose that the entire loop, including the measuring device, is composed of the same material, then it is clear that the EMF we are seeking to measure will be exactly cancelled by an equal and opposite EMF in the measurement leg of the loop. The solution is to use dissimilar materials and to measure the difference between the thermally generated EMFs. This differential EMF is called the Seebeck EMF, and this is the EMF on which thermoelectric temperature measurements are based.

Other electrical and thermoelectric effects occur that may need to be considered as possible sources of measurement error, or alternatively the effects may in some cases form applicable sources of heating and cooling in their own right.

If an electric current is allowed or caused to flow around a loop such as that in Fig. 3.1 then of course Joule (i^2R) heating will occur. In addition, it will be found that the current flowing through the junctions causes temperature increase at one junction and temperature decrease at the other – this is called the Peltier effect.

A final phenomenon is that a current travelling in a conductor that is subject to a temperature gradient causes either an evolution or an absorption of heat depending on the direction of the current and the temperature gradient. This effect, which is very small, is called the Thompson effect.

In summary, the Seebeck effect causes a thermoelectric EMF to appear between the differentially heated junctions of a loop made from two dissimilar conductors. If current flows around such a loop, the Peltier effect produces heating and cooling, respectively, at the two junctions. (The Peltier effect is particularly useful as a source of electrical cooling.) Joule and Thompson effects are additional phenomena that complete the description of behaviour in the thermocouple circuit.

3.3 Practical laws of thermoelectricity

3.3.1 Law of homogeneous circuits

A thermoelectric current cannot be sustained in a circuit composed of a single material. (If any such current is observed it should be taken as an indication of non-homogeneity in the material.)

3.3.2 The law of intermediate metals

The algebraic sum of thermally generated EMFs in a circuit composed of any number of different materials is zero provided that the circuit is at a uniform temperature.

Implication 1: A third material, inserted into a thermocouple circuit and kept at a uniform temperature along its whole length will have no effect on the overall EMF of the circuit.

Implication 2: Let A, B, C be three different metals and define e_{AB} as the thermoelectric EMF between metals A and B:

$$\text{Then} \quad e_{AB} = e_{AC} + e_{CB} \tag{3.1}$$

and knowledge of thermoelectric EMFs against a reference metal allows the EMF of any combination to be calculated.

3.3.3 Law of intermediate temperatures

Suppose that two dissimilar metals produce thermoelectric EMFs as shown below:

EMF	Junction temperatures	
e_{12}	θ_1	θ_2
e_{23}	θ_2	θ_3
Then the following relation holds:		
e_{13}	θ_1	θ_3
$= e_{12} + e_{23}$		

Implication: Additional junctions (electrical connections) must necessarily be made at the head of a thermocouple. The thermocouple head may be at an arbitrary temperature, but provided that the wires used to carry the thermocouple signals have an identical composition (or identical thermoelectric properties) to the thermocouple wires, there will be no extraneous EMF introduced into the overall circuit.

Such wires matching the thermoelectric properties of the thermocouple are called *compensating cables* (see Section 3.7). It is clearly a matter of practical importance that compensating cables are connected to the thermocouple head with correct polarity.

3.4 Cold-junction compensation

A thermocouple measures the temperature difference between the hot junction and a cold junction that will usually be located within the measuring instrument. Since the measuring instrument will be at a varying ambient temperature, so-called *cold-junction compensation* is needed to ensure that correct absolute temperature measurements are obtained.

A number of possible electromechanical (e.g. bimetallic-strip- driven) or electronic devices are available for cold-junction compensation, and they can be incorporated with other signal-processing components to achieve combined cold-junction compensation, break protection, linearisation and signal amplification.

For high-accuracy applications, it will often be preferable to site the cold junction in a special-purpose precisely controlled oven. Of course, precise control of this oven is then essential, and if such control is achieved using a control loop containing a thermocouple, the question arises: how is the cold junction of this thermocouple to be compensated. The answer is that a precisely controlled oven for cold-junction compensation can be obtained, but at a high cost. However, once the oven is available, it may be used to house a large number (up to 50) of cold functions – always provided that the original thermocouples, whose cold-junctions are to be compensated, are sufficiently closely geographically clustered.

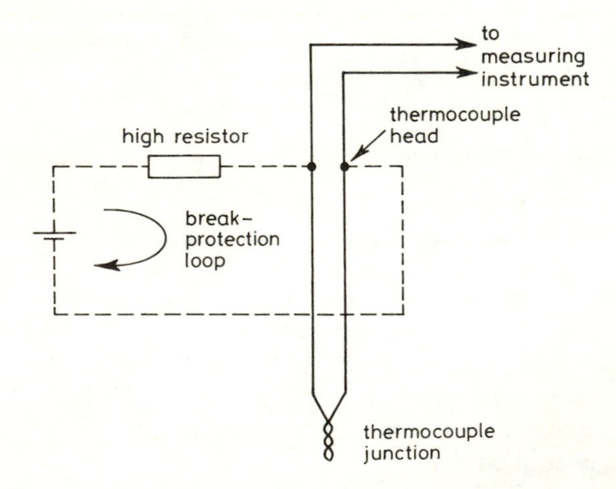

Fig. 3.3 *Simple thermocouple break-protection circuit*
The measurement is driven up-scale when the thermocouple junction becomes open circuit

3.5 Break protection

Thermocouples often fail by becoming open-circuit, and a dangerous situation can arise in that a low-temperature will then be indicated and temperature controllers will, incorrectly, open fuel valves fully.

To avoid this problem it is necessary to include some means by which the system fails safe when the thermocouple becomes open circuit. One simple means is shown in Fig. 3.3. Under normal conditions the break-protection loop causes a negligible voltage to be developed across the thermocouple junction. When the junction becomes open circuit a voltage appears across the junction of such a polarity as to drive the indicated temperature to a high setting.

3.6 Materials used for thermocouples

The ideal characteristics of the metals constituting a thermocouple would be:

- high sensitivity and linearity
- time-invariant behaviour
- no device-to-device variability
- physical strength and stability at high temperatures
- resistance to effects of corrosive agents
- low cost

3.6.1 Base-metal thermocouples

So called *base metal thermocouples* use relatively low-cost conductors and are adequate for most industrial processes operating below about 900°C. Base-metal thermocouples are of relatively low cost and high sensitivity, but they have a nonlinear output and their long-term stability makes them unsuitable for applications demanding highest possible accuracy.

Popular base metal thermocouple materials include:

Type K: nickel–chromium (Chromel)	v. nickel–aluminium (Alumel)
Type J: iron	v. copper–nickel
Type T: copper	v. copper–nickel (Constantan)

The output voltages of these thermocouples against temperature are shown in Fig. 3.4.

3.6.2 Platinum/platinum alloy thermocouples

Platinum is used because it can be obtained in a high state of purity and because it has high resistance to most forms of chemical attack combined with good mechanical properties at high temperature.

The type S thermocouple uses one lead of pure platinum and one of 90% platinum, 10% rhodium. This thermocouple is used to define the international temperature scale over the range 630·5° to 1063°C. Thermocouples for general use are constructed from materials of 99·99% purity, whereas thermocouples to be used as calibration standards utilise materials of 99·999% purity.

The type R thermocouple uses one lead of pure platinum and one of 87% platinum, 13% rhodium.

Platinum/platinum–rhodium thermocouples are used widely to measure temperatures in the range 400°–1500°C. Despite the relatively good chemical resistance of platinum/platinum–rhodium, thermocouples do lose accuracy or fail mechanically because of absorption of contaminants from furnace gases, embrittlement of the conductors, evaporation of platinum and rhodium and because of diffusion of rhodium through the junction.

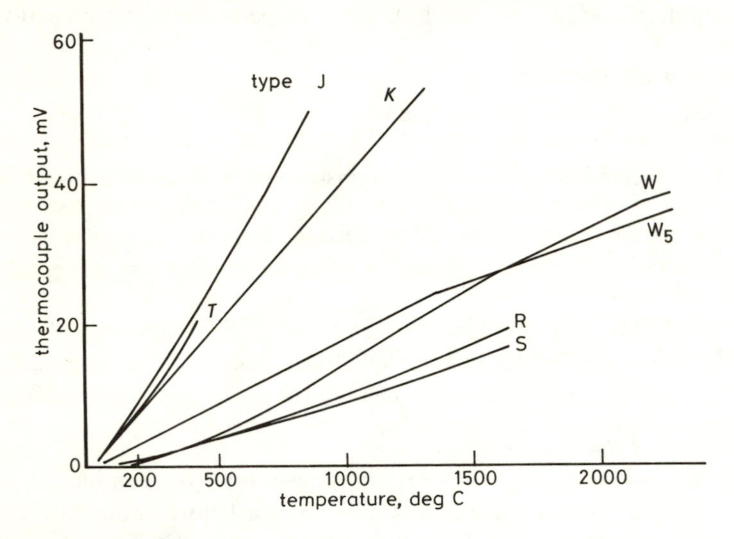

Fig. 3.4 ***EMF/temperature for selected thermocouple materials***
Type J: Iron v. Constantan
Type T: Copper v. Constantan
Type K: Chromel v. Alumel
Type W: Tungsten v. tungsten/26% rhenium
Type W_5: Tungsten/5% rhenium v. tungsten/26% rhenium
Type R: Platinum v. platinum/13% rhodium
Type S: Platinum v. platinum/10% rhodium

The characteristic curves for types R and S thermocouples are given in Fig. 3.4.

3.6.3 High-temperature thermocouples

Temperatures up to 2600°C may be measured by the thermocouples:

Type W: tungsten v. tungsten/25% rhenium
Type W5: tungsten/5% rhenium v. tungsten/26% rhenium

The output voltages of these two types are shown in Fig. 3.4.

3.7 Compensating cables

If ordinary copper conductors are connected to the head of a thermocouple, the junctions between dissimilar metals so formed generate additional thermal EMFs that may give rise to erroneous measurement.

To avoid such problems, wires (*compensating cable*) having the same thermoelectric properties as those of the thermocouple are used to bring the signals from the thermocouple head to the temperature-monitoring equipment. For base-metal thermocouples, compensating cable usually uses the same materials as those in the thermocouple, suitably stranded to allow flexibility.

3.8 Thermocouple sheaths

Thermocouple leads are kept separate by means of hollow insulators through which they are threaded. The insulated leads are then usually mounted in a refractory sleeve, which in turn (for industrial use) is mounted inside a metal sheath. The metal sheath needs to be chosen to match the conditions (temperature, presence of corrosive gases or liquids etc.) that are expected in the application.

Sheath materials range from mild steel to special high-performance alloys for arduous applications. Special sheath materials have been developed for particular applications, such as measurement of temperature of molten salt baths; manufacturers' literature should be consulted.

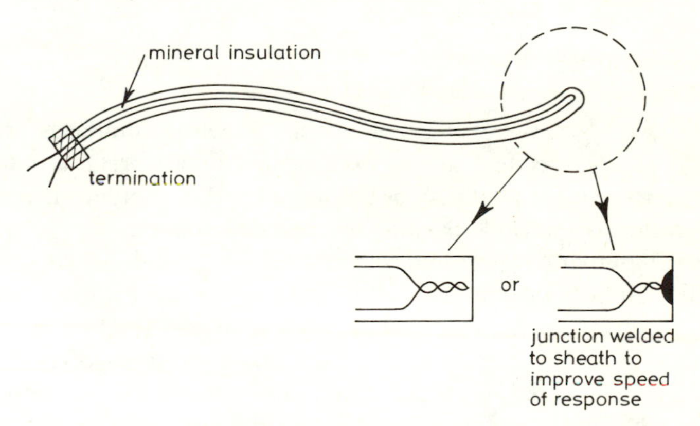

Fig. 3.5 *Mineral-insulated thermocouple*

3.9 Mineral-insulated thermocouples

The term *mineral-insulated thermocouple* is generally taken to define a class of small-diameter (typically 3 mm) flexible probes (Fig. 3.5).

They are usually of low cost, and because of their small diameter and flexibility they can be installed in difficult locations. As an example, Fig. 3.6 shows how the temperature inside a workpiece inside a hot oven may be monitored using a mineral-insulated thermocouple.

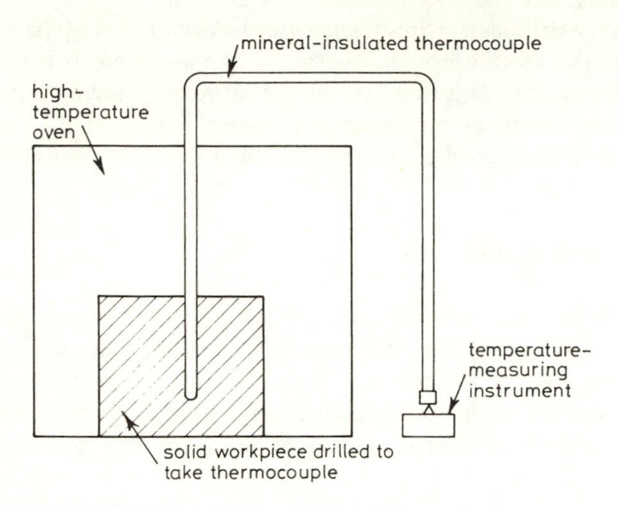

Fig. 3.6 *Illustrating the use of a mineral-insulated thermocouple to obtain the internal temperature of a workpiece*
The thermocouple successfully passes through the intermediate high-temperature region

3.10 Calibration of thermocouples

Calibration may be carried out either against absolute standards established locally (such as are specified in the fixed points table) or, alternatively, the thermocouple to be calibrated may be compared with a standard thermocouple whose calibration is traceable back to the primary standard.

The second (comparison) method is the more widely used. Usually, a thermocouple that is already installed and working will be checked at a single temperature by inserting the standard thermocouple so that the two hot junctions are as close as possible to each other. The output voltages of both thermocouples are then measured by an accurate millivoltmeter whose temperature is measured by a built-in thermometer to allow correction to be made for ambient temperature.

Thermocouple tolerance specifications vary according to thermocouple type and temperature range, and the references and makers' tables should be consulted for details. However, as a guide, a tolerance of $\pm 1°C$ is typical even up to temperatures of 1200°C.

3.11 Siting of thermocouples

Thermocouples must be mounted where they will measure representative temperatures in positions that will allow an economically viable life for the sensor before it has to be replaced.

In addition to the obvious precaution of avoiding hot and cold spots (for instance, locations where flames will impinge on the thermocouple in fuel-fired furnaces), a sufficient length of immersion must be allowed. However, very long horizontally immersed thermocouple sections will be prone to sag, with the effect being dependent on temperature, sheath diameter and sheath material.

3.12 Special thermocouple probes

To allow temperature measurement in particular industrial applications, specialised probes have been developed. They include:

Fast-acting probes for air-temperature measurement (these can be surrounded by blackened spheres to measure radiation, but if this is done, the time constant is very considerably increased).
Hyperdermic probes – for insertion into rubber etc.
Chisel probes – for hammering into frozen food etc.
Pipe-clamp probes – for encircling pipes
Molten-metal probes.

3.13 Dynamic response of thermocouples

When a thermocouple at temperature θ_1 is suddenly immersed in a fluid at temperature θ_2, the subsequent time response is approximately exponential, with the time constant being dependent on the diameter of the thermocouple wires, the material, thickness and type of construction of the sheath, the nature of the fluid and the state of motion of the fluid.

As an illustration, Fig. 3.7 is included. It shows how the physical construction of a thermocouple affects the response time in a particular set of conditions.

When a given thermocouple is immersed in a moving fluid, the response time is highly dependent on both the nature and the velocity of the fluid. Let the time/constant of the thermocouple response be T seconds and the velocity of the fluid in which it is immersed by v metre/second, then, very approximately, it is found that

$$T = T_0 e^{av} \qquad (3.2)$$

in which typical values for an industrial thermocouple are:

for water; $T_0 = 10, \quad a = -1.5$

for air; $T_0 = 200, \quad a = -1$

Thus we see that the quoted thermocouple in still air has a time constant of 200 s (whereas in a 2 m/s air stream this falls to 73 s.) The same thermocouple in still water has a time constant of 10 s, (falling to 2·2 s when the water stream has a velocity of 1 m/s.)

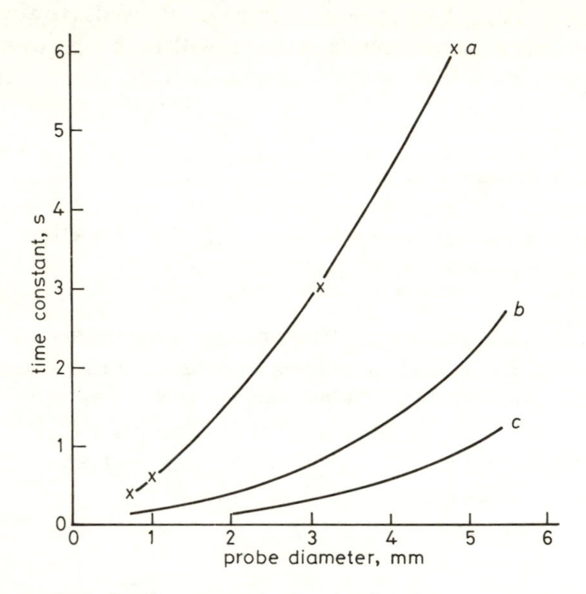

Fig. 3.7 *Time constants for thermocouples in stirred water*
a Inside a stainless-steel closed-end probe
b Response of mineral-insulated thermocouple
c Response of mineral-insulated thermocouple with junction welded to sheath

Both the construction of the thermocouple and the characteristics and velocity of the fluid into which it is immersed have a major effect on the time constant of the resulting measurement circuit. Care must be taken, in the context of each application, that the overall control performance is not degraded by an excessive measurement time constant. For more background on this topic see Baker *et al.* (undated).

3.14 Direct head-mounting thermocouple transmitters

A thermocouple transmitter is a device that converts the low-level thermocouple signal into (typically) a 4–20 mA current signal. Direct head-mounting transmitters are small enough to fit inside the head of a standard thermocouple. Thus there is no need to use compensating cables. The use of transmitters also increases the immunity of the system to radio interference.

Available devices include break protection and cold-junction compensation within their circuitry. Similar head-mounting transmitters are available for resistance thermometers.

3.15 References and further reading

ASTM (USA): 'Manual on the use of thermocouples in temperature measurement'

BAKER, H. D., RYDER, E. A., and BAKER, N. H.: 'Temperature measurement in engineering' Vols. I and II (Omega Press, USA) (undated)

BARBER, R., and BROWN, M.E. (1979): 'Thermocouples above 600°C: Infra-red thermometers and optical pyrometers'. Measurement & Control, 12, May

BARNARD, R. D. (1972): 'Thermoelectricity in metals and alloys' (Taylor & Francis, London)

CANNON, C. P. 1982: '2200C thermocouples for nuclear fuel centreline temperature measurements' *in* SCHOOLEY, J. F. (Ed.): 'Temperature: its measurement and control in science and industry'. Vol. 5 (American Institute of Physics)

EVANS, J. P. (1982): 'Experiences with high temperature platinum thermometers' *in* SCHOOLEY, J. F. (Ed.): 'Temperature: its measurement and control in science and industry'. Vol. 5 (American Institute of Physics)

NBS (USA): 'Thermocouple reference tables'. NBS Monograph 125

POLLOCK, D. D.: 'The theory and properties of thermocouple elements', Omega Press (USA)

REED, R. P. (1982): 'Thermoelectric thermometry: A functional model' *in* SCHOOLEY, J. F. (Ed.): 'Temperature: its measurement and control in science and industry' Vol. 5 (American Institute of Physics)

Chapter 4

Resistance thermometers and thermistors

4.1 Principles

The electrical resistance of almost every material varies with temperature. The resistance of metal conductors generally increases with temperature in a repeatable manner, but at a low sensitivity. Measuring sensors exploiting this phenomenon to measure temperature are called *resistance thermometers*.

The resistance of semiconductor materials with temperature varies in a more complex manner – usually, but not always, resistance falls with rising temperature. Certain semiconductor materials have a very high rate of change of resistance with temperature, making them very attractive for use as temperature sensors if problems such as stability and device-to-device variability can be overcome. Temperature sensors based on resistance changes in semiconductor materials are called thermistors.

In general, resistance thermometers find application in situations demanding high accuracy. Thermistors are applied where high sensor output and low cost are required.

4.2 Resistance thermometers

A resistance thermometer (RTD) is made up from resistance wire wound on a suitable former. The wire must have a known characteristic relating resistance with temperature, and this characteristic must not change with time. The resistance element is placed inside a sheath to protect it against physical damage or chemical attack. The sheath is sometimes packed with an electrically insulating powder to protect against vibration and to improve heat conduction from sheath to resistance element.

Pure metal wires are usually used to form the resistance elements. Nickel, copper and platinum are the most widely used materials and their characteristics are as follows:

Material	Temperature range	Element resistance at	
		0°C	100°C
Nickel	−60 to 180°C	100 Ω	152 Ω
Copper	−30 to 220°C	100 Ω	139 Ω
Platinum	−200 to 850°C	100 Ω	136 Ω

The advantage of nickel's high coefficient of change is partly offset by its markedly nonlinear characteristics. Copper wire has a very low resistance compared with the other two materials, so that a long length of thin wire is required to form an element of reasonable resistance value. Platinum remains the best material for many applications.

The upper usable temperature is often fixed by the type of former on which the resistance wire is wound. Ceramic formers have the highest temperature specification (up to about 850°C) but such formers are liable to fracture early, particularly if they are subject to vibration.

4.2.1 Tolerances on commercially available resistance thermometers

The tolerance specification for Class A 100 Ω platinum-resistance thermometers has been laid down in DIN 43760/BSI1904 documents over the range −200 to 650°C and may be summarised in the equation

$$\text{tolerance (°C)} = \pm(0{\cdot}15 + 0{\cdot}002|\theta|) \qquad (4.1)$$

where θ is the operating temperature, deg C.

Some advantages of resistance thermometers over thermocouples are:

- Resistance-thermometers measure temperature directly, with no reference temperature being involved.
- The physical laws governing their behaviour are less complex than for thermocouples. This implies a higher potential mesurement accuracy.
- Long resistance thermometer elements may be used to measure average temperatures, e.g. in fuel storage tanks.

Because of these advantages, the application of resistance thermometers is increasing, although the upper usable temperature (about 800°C) is still well below that for thermocouples.

Over the years the platinum-resistance thermometer has been steadily improved, and the development continues, with the aim of allowing higher maximum operating temperatures to be achieved.

A disadvantage of resistance thermometers is their relatively small change of resistance with temperature, making high amplification necessary. Even more important: inevitable changes in the temperature (and hence in the resistance) of connecting wires may be interpreted incorrectly as changes in the measured temperature. To minimise such sources of error, various compensating bridge circuits have been devised – Fig. 4.1 shows one such bridge circuit.

Another development in resistance thermometry is that of miniature (down to 2 × 2 × 1 mm) elements made by depositing platinum film on a silicon base. These devices are of low cost, small size, rapid response and high accuracy, and they may be integrated into a hybrid chip that will provide the necessary signal conditioning and amplification required for direct digital read out of temperature. The low-cost/small-physical-size combination is opening up new application areas for the devices, e.g. in the temperature control of domestic cooking hobs.

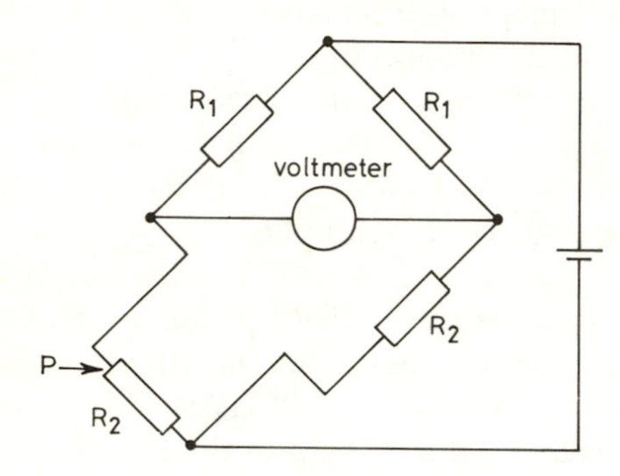

Fig. 4.1 *Three-wire bridge for measurement of changes in the resistance thermometer P*
The arrangement compensates for the effect of lead resistance on the measurement

4.2.2 Bridge arrangements for the measurement of resistance

Several different bridge arrangements are used to measure the resistance of RTDs. Most are designed to compensate for the errors that could arise from changes in lead resistance. One bridge arrangement is shown in Fig. 4.1. More detailed information may be found in manufacturers' catalogues.

4.3 Thermistors

Thermistors are resistors that have a high rate of change of resistance with temperature. They consist of sintered beads of semiconductor material encapsulated in a glassy compound.

In the past, thermistors acquired a poor reputation based on poor stability and device-to-device variability, resulting in resistance tolerances of ±10%, or worse. Present-day thermistors have tight tolerances, enabling temperature measurements to better than ±0·2 deg C in the −30° to +100°C range to be made. Thermistors are available to span the range −100° to +300°C.

Most thermistors have a negative coefficient relating resistance with temperature, but positive-temperature-coefficient (PTC) thermistors are also available and they are sometimes used because of their greater sensitivity. PTC thermistors have a positive coefficient only over a limited temperature range – outside this range their coefficient is negative. PTC themselves therefore tend to be applied in those situations where control to a fixed setpoint is required.

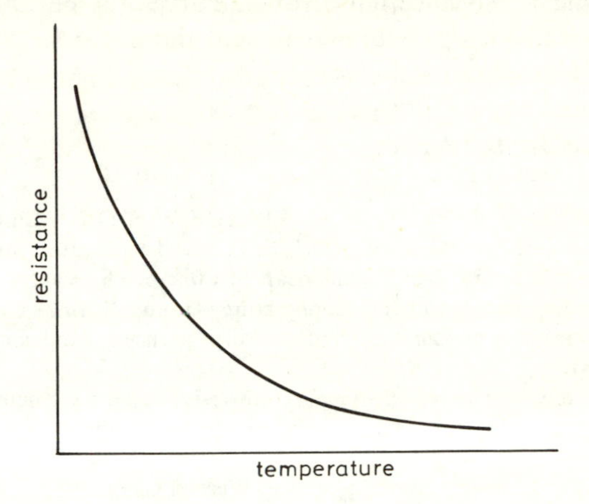

Fig. 4.2 *Typical resistance/temperature curve for a negative-temperature-coefficient thermistor*

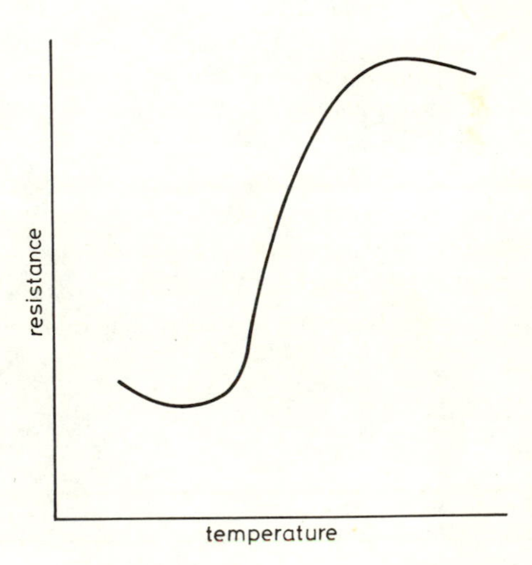

Fig. 4.3 *Typical resistance/temperature curve for a positive-temperature-coefficient thermistor*

Advantages of thermistors are:

- High sensitivity
- Small physical size
- Resistance at ambient temperature may be specified anywhere in the range 0·1–300 kΩ

These factors allow the design of a wide range of low-cost application probes whose signals require much less amplification than do signals from thermocouples or resistance thermometers. The disadvantages of thermistors are their limited temperature range and their modest upper usable temperature. Figs. 4.2 and 4.3 show typical resistance/temperature curves for negative-temperature-coefficient and positive-temperature-coefficient thermistors.

4.4 References and further reading

BERGER, R. L., BALKO, B., CLEM, T. R., and FRIAUF, W. S. (1982): 'Fast thermistor sensors for rapid reaction studies' *in* SCHOOLEY, J. F. (Ed.): 'Temperature: its measurement and control in science and industry'. Vol. 6 (American Institute of Physics)

EVANS, J. P. (1982): 'Experiences with high temperature platinum thermometers' *in* SCHOOLEY, J. F. (Ed.): 'Temperature: its measurement and control in science and industry'. Vol. 6 (American Institute of Physics)

McALLAN, J. V. (1982): 'Practical high temperature resistance thermometry' *in* SCHOOLEY, J. F. (Ed.): 'Temperature: its measurement and control in science and industry'. Vol. 5 (American Institute of Physics)

SEARS, F. W. (1982): 'University physics' (Addison Wesley) Chap. 28

Chapter 5

Radiation thermometry

5.1 Principles

Hot objects above a temperature of about 500°C emit visible light, but by far the greater proportion of radiation is emitted in the infra-red part of the spectrum. Even at ordinary temperatures near ambient, the emission of infra-red radiation is sufficient to form a basis for accurate measurement by radiation techniques.

Consider a solid body at any temperature T. It is found that the body radiates energy according to a pattern that depends only on two factors:

(i) Temperature T
(ii) Condition of the body as quantified by a coefficient E called the emissivity of the body.

The energy H radiated from a surface of area A in unit time is given by

$$H = AE\sigma T^4 \qquad (5.1)$$

where σ is the Stefan–Boltzmann constant.
Radiation thermometry depends on measuring radiated energy and, after correcting for emissivity, determining the temperature from the physical laws governing radiation.

Planck's law describes the distribution of energy from a perfect radiator of maximum possible emissivity (emissivity coefficient unity) according to the equation

$$J(\lambda) = \frac{c_1 \lambda^{-5}}{\exp(c_2/\lambda T)} \qquad (5.2)$$

where λ is the wavelength of the radiation, c_1, c_2 are constants and J represents the intensity of radiation emitted at wavelength λ per unit area of the body.

Fig. 5.1 shows the form of the spectral-energy distribution as described by Planck's law. Let λ_m be that wavelength where maximum energy intensity occurs; then it is clear from the Figure that λ_m shifts to the left as temperature

T rises. This shift of the wavelength of maximum energy intensity is quantified in *Wien's displacement law*:

$$\lambda_m T \;=\; \text{constant} \tag{5.3}$$

A body that has emission coefficient unity emits the maximum possible radiation and is called a *black body*. All actual objects have emissivities of less than unity so that a black body is a non-attainable idealisation. (However, radiation emanating from a small hole in a heated enclosure approaches close to black-body conditions, and such an arrangement is therefore used to calibrate radiation thermometers against standard temperatures.)

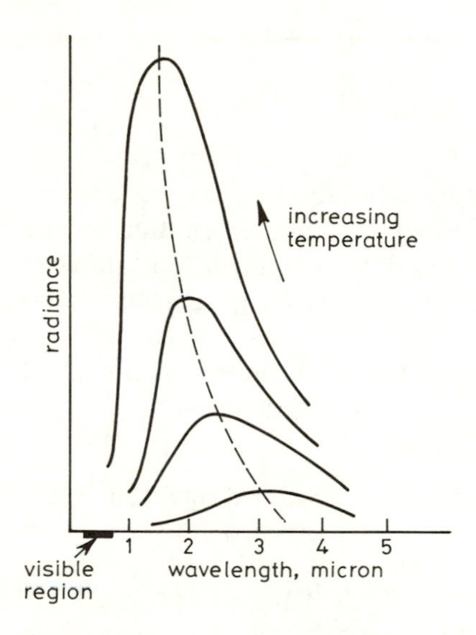

Fig. 5.1 *Distribution of radiance in black-body radiation*

A *gray body* is a body of emissivity coefficient $E < 1$ whose radiation pattern is given by multiplying the right-hand side of eqn. 5.3 by the emissivity coefficient; i.e. the radiance of a gray body is given by

$$J(\lambda) \;=\; \frac{EC_1\lambda^{-5}}{\exp\,(c_2/\lambda T)} \tag{5.4}$$

where E $(0 < E < 1)$ is the emissivity coefficient.
A *non-gray body* has an emissivity factor $E(\lambda)$ that varies with wavelength. Non-gray bodies do not emit energy according to the pattern given by eqn. 5.4. (However, at no wavelength can the intensity exceed that described by eqn. 5.4.)

Fig. 5.1 illustrates black-body radiation at different temperatures as described by eqn. 5.4. Fig. 5.2 illustrates the radiation from a black body, a gray body and a non-gray body all at the same temperature T. Clearly, eqn. 5.4 is of little value in establishing a quantitative relation between temperature and radiation intensity in the case of a non-gray body.

Most solids and liquids are gray bodies to a reasonable approximation. Some emissivities are given in Table 5.1 for a few selected materials. Notice from the Table that the emissivity depends on both composition and surface condition. This fact introduces uncertainty into measurement of the temperature of metals

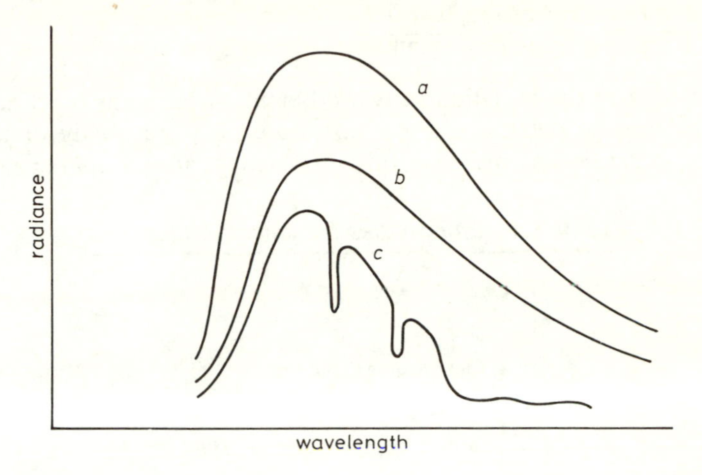

Fig. 5.2 *Distribution of radiance from*
a Black body
b Gray body
c Non-gray body
All at the same temperature

at high temperatures, since, although the composition of a metal may be known accurately, the surface will often be covered with a non-homogeneous layer of oxides whose composition and physical condition cannot be accurately known. To make a temperature measurement of a gray body using radiation techniques two basic approaches are possible:

(i) To measure the radiation intensity at a single chosen wavelength λ. (In practice of course, within a narrow band centred on λ.)
(ii) To measure, by some means, the area under the total radiation intensity curve (Fig. 5.1).

In both cases, combined with a knowledge of the appropriate emissivity coefficient, the information is sufficient to allow temperature to be calculated.

(iii) (A variant of (i) that uses measurements of radiation intensity at two chosen wavelengths λ_1, λ_2 to determine temperature of a body of unknown emissivity).

Referring to eqn. 5.4 and substituting the values T, λ_1, λ_2 to obtain, assuming that the emissivity E is constant,

$$J(\lambda_1) \quad = \quad \frac{Ec_1\lambda_1^{-5}}{\exp\,(c_2/\lambda_1 T)}$$

$$J(\lambda_2) = \frac{Ec_1\lambda_2^{-5}}{\exp(c_2/\lambda_2 T)}$$

and

$$\frac{J(\lambda_1)}{J(\lambda_2)} = \frac{\lambda_1^{-5}\exp(c_1/\lambda_2 T)}{\lambda_2^{-5}\exp(c_2/\lambda_1 T)} \tag{5.5}$$

Thus the ratio of the radiation at two different wavelengths may, at least in theory, be used to determine the temperature of a gray body of unknown emissivity. A device making use of this approach is called a *ratio* or *two-colour pyrometer*.

Table 5.1 *Emissivities for selected materials*

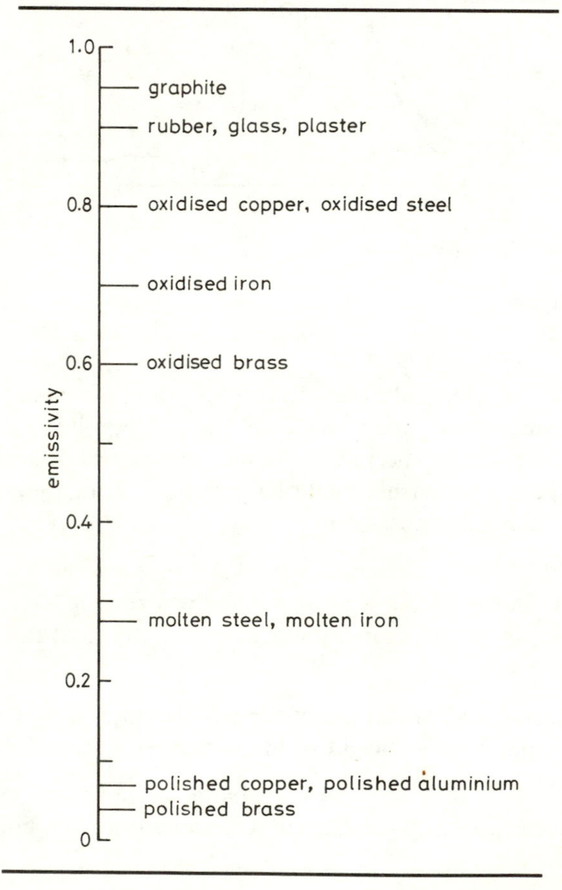

5.2 Choice of operating waveband

Early radiation pyrometers often used a very wide spectral band to obtain sufficient incident radiation onto a thermal detector. Current practice is to use a relatively narrow band centred on the lowest possible wavelength, since this allows the best possible sensitivity – the greatest values of $dR/d\theta$, where R is radiance, are clearly obtained at the low wavelengths.

However, as the temperature to be measured becomes lower, it is necessary

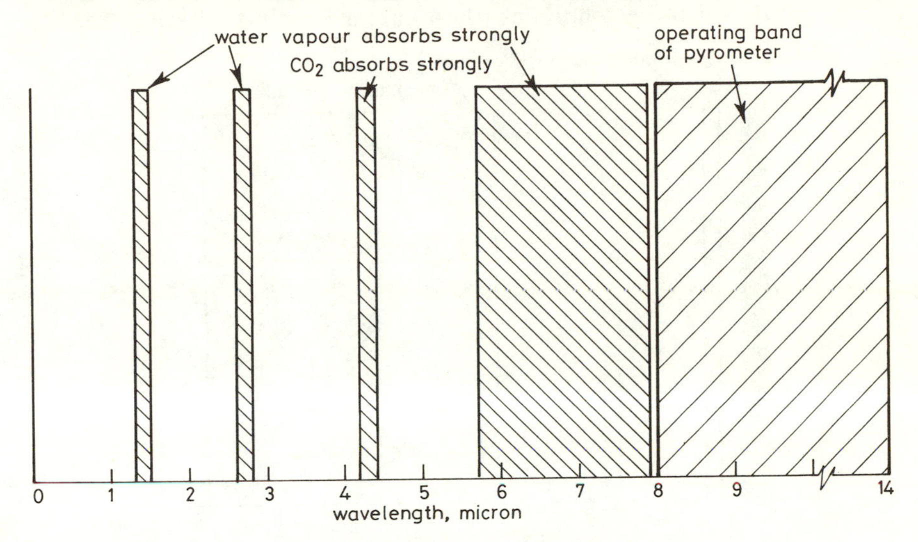

Fig. 5.3 *Wide-temperature-range pyrometer, −50°C to 1000°C*

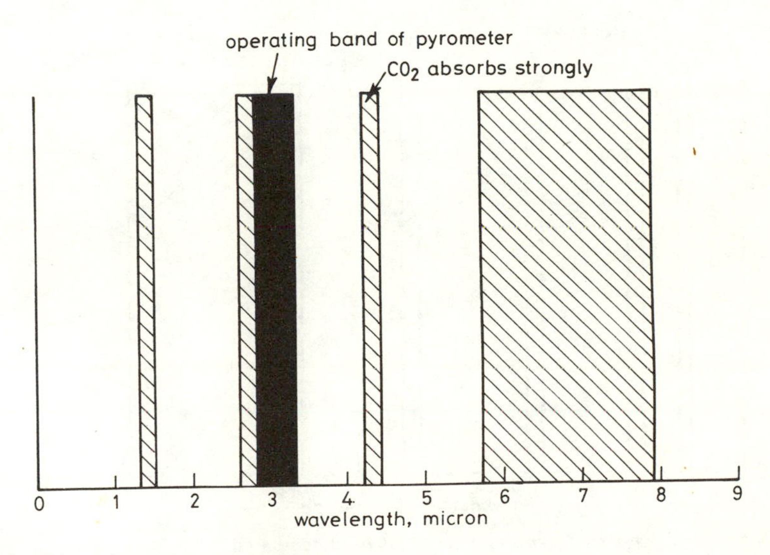

Fig. 5.4 *General-purpose pyrometer*

to move to longer wavelengths in order to obtain sufficient radiation to drive the detector. A number of other points have to be considered and, overall, the following factors have to be taken into account:

(i) Temperature range of the target
(ii) Nature, size, uniformity and location of the target
(iii) Environment along the proposed sight path, including the presence of intermittent obstructions, smoke, hot gases and water vapour
(iv) Extent to which the conditions (i)–(iii) are narrowly and well defined.

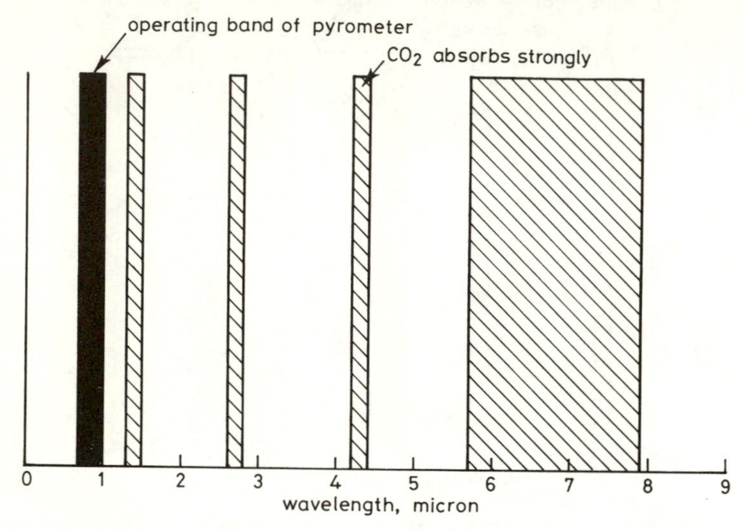

Fig. 5.5 *General-purpose high-temperature (>700°C) pyrometer*

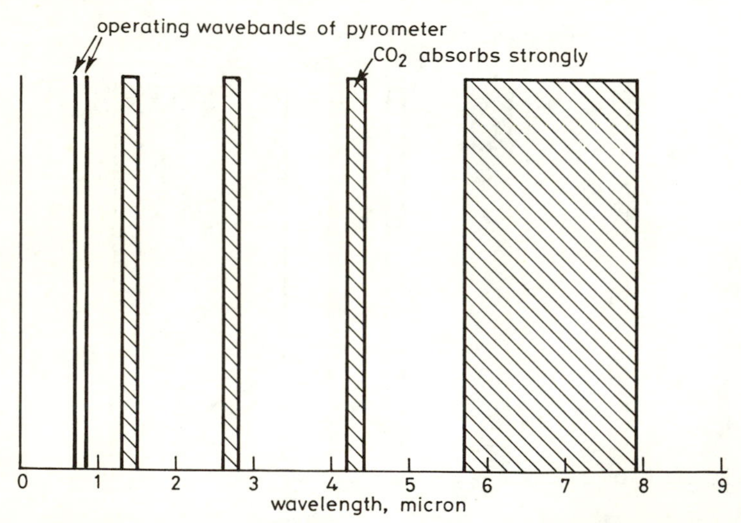

Fig. 5.6 *Ratio pyrometer for measurement of high temperatures where the target may be partially obscured*

In general, the operating waveband will be chosen to yield maximum sensitivity to temperature variation over the temperature range of interest, while minimising the errors due to uncertainties in target emissivity and due to radiation absorption along the sight path.

A general-purpose radiation pyrometer, able to operate in a wide variety of situations, will necessarily be a compromise compared with a device that has been closely specified for a specific application.

Figs. 5.3–5.8 show, respectively, typical operating wavebands for pyrometers

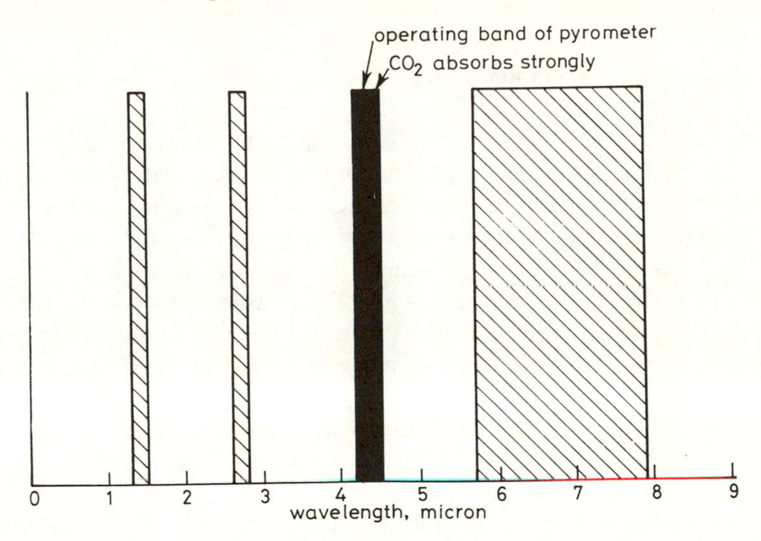

Fig. 5.7 *Flame-temperature pyrometer*
Operating band includes region where CO_2 absorbs/radiates strongly

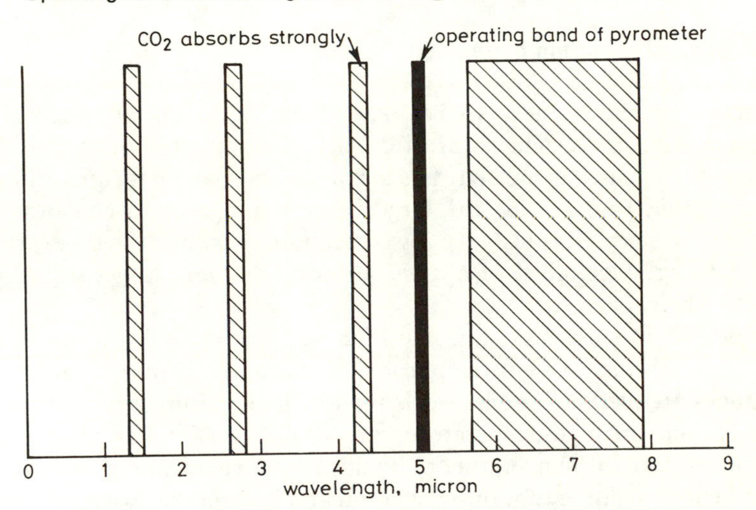

Fig. 5.8 *Pyrometer for measuring temperature of glass surface*
Glass is opaque at the chosen wavelength

for: wide temperature range, general purpose, high temperature, ratio method of measurement, flame-temperature measurement and glass-surface-temperature measurement. Collectively the Figures show how measurements through furnace atmospheres are located at regions in the waveband away from gas-absorption effects, with shorter wavelengths being used as the measured temperature increases. Special measurements, like those for flame or glass-surface-temperature are located at wavelengths where the phenomena of interest are known to emit strongly.

5.3 Transmission of radiation through gases

Gases are non-gray bodies. Their radiation spectrum does not follow the form given by eqn. 5.2, but rather they emit or absorb only within certain narrow wavebands as shown approximately for CO_2 and water vapour in Fig. 5.9. This fact makes hot-gas-radiation thermometry an area calling for particular expertise. It also means that, where the temperature of a hot body has to be measured by radiation techniques, through a layer of hot gas (a fairly common industrial situation), interfering radiation effects from the gas will be difficult to calculate by theoretical methods. It is therefore essential to confine such measurements to precisely known bandwidths away from regions in the spectrum where interfering phenomena occur. Otherwise the measurements obtained will be affected in both accuracy and repeatability by the condition of the atmosphere between target and detector.

A ratio pyrometer will be particularly badly affected if one of the measuring wavelengths is attenuated more than the other by atmospheric absorption.

5.4 Sighting of radiation pyrometers

Radiation pyrometers need to be focused on to targets of widely differing maximum size across sight paths of differing lengths. It is important that the object to be measured completely fills the target zone, and it is equally important that radiation from outside the target zone is not able to enter the optical measuring system, for instance by reflection. Manufacturers of radiation pyrometers produce tables of options allowing different target sizes and path lengths to be catered for.

Where a sight path passes through, for instance, a smoky atmosphere, it is often advantageous to sight the pyrometer down an air-purged sighting tube. Such tubes are made of special steels for low-temperature applications and of silicon carbide for high-temperature (greater than 1000°C) applications.

For measuring the temperature of liquids it is often advantageous to immerse a closed-ended sighting tube into the liquid and to sight the radiation pyrometer down the tube with the closed end as its target.

5.5 Optical system for a radiation pyrometer

Optical glass is opaque to long-wavelength radiation and lenses must therefore be constructed of other materials, such as calcium fluoride (see Fig. 5.10).

Alternatively, concave mirrors may be used to concentrate radiation onto the radiation detector.

In the *total radiation pyrometer*, the radiation within a sharply defined solid angle is focused by an optical system onto a blackened thermocouple inside

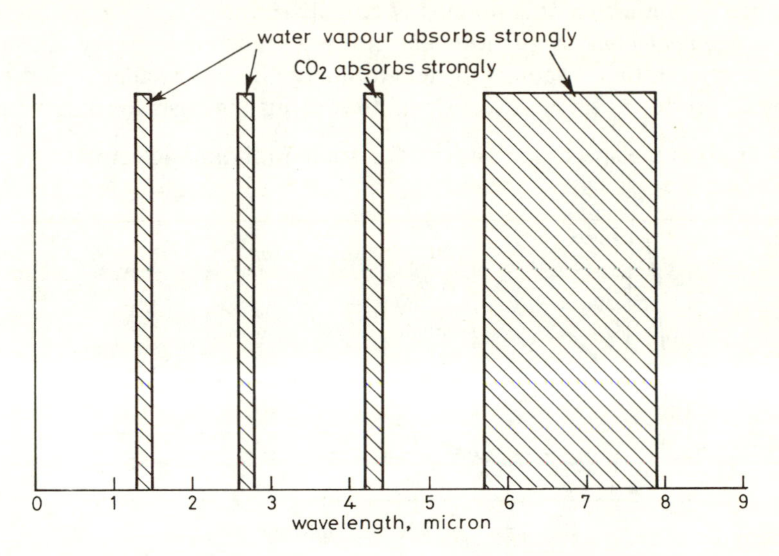

Fig. 5.9 *Regions of the infra-red spectrum where atmospheric water vapour and CO_2 absorb radiation strongly*

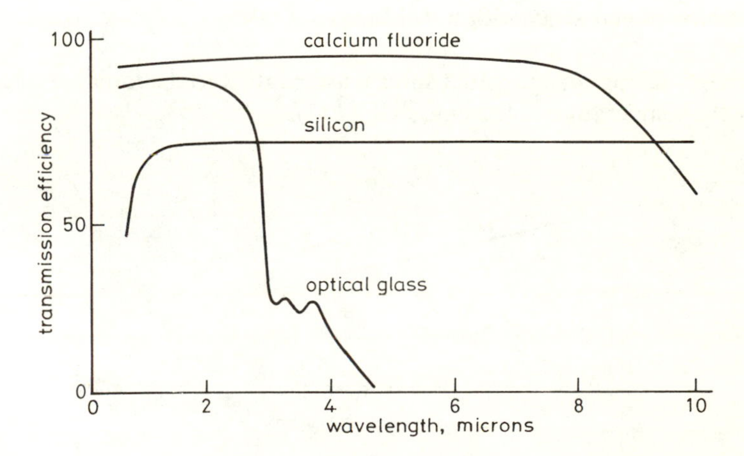

Fig. 5.10 *Transmission characteristics of some alternative optical materials*

either an evacuated bulb (for maximum sensitivity) or a gas-filled bulb (for maximum speed of response). Subject to modification, to take into account the frequency-dependent attenuation in the optical system, the instrument allows the total radiation per unit area from the target body to be determined, and this allows the temperature to be calculated.

5.6 Radiation detectors

Two types of radiation detectors are used:

(i) *Thermal detectors*: A thermal detector is a blackened thermocouple or resistance thermometer. It is nominally equally sensitive to all wavelengths.
(ii) *Quantum detectors*: A quantum detector is a semiconductor crystal, such as a silicon crystal, that responds to radiant energy in a particular region of the spectrum, producing an electrical signal that forms the basis for measurement.

Fig. 5.11 gives the spectral sensitivity for some radiation detectors.

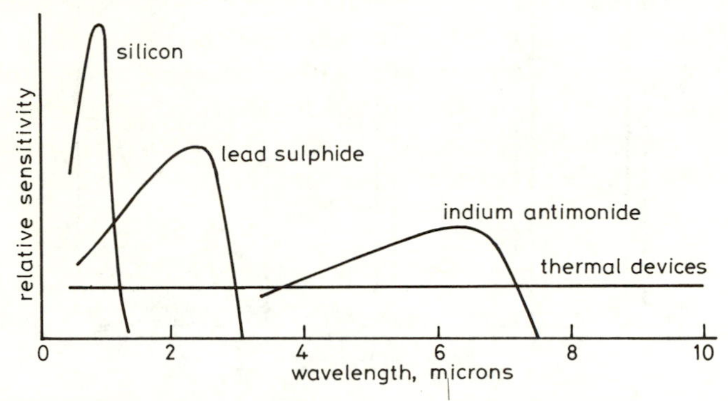

Fig. 5.11 *Spectral sensitivity of some detectors*

5.7 Variation of emissivity with wavelength

The ratio of radiance of a target at some wavelength λ to the radiance of a black body at the temperature is the emissivity $E(\lambda)$.

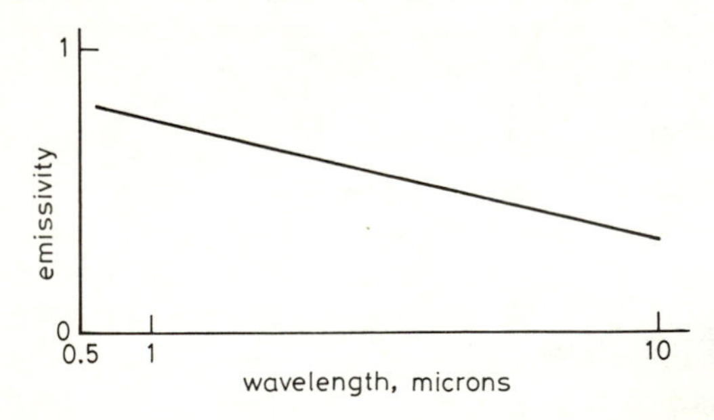

Fig. 5.12 *Emissivity of most metals is higher at short wavelengths. Typical emissivity variation for a metal*
In general, emissivity is also an (increasing) function of temperature

For a gray body, $E(\lambda)$ is not a function of λ and we can talk of the emissivity E. However, many practical targets are not true gray bodies, and it will be preferable to choose for measurement a part of the spectrum where the emissivity is high and repeatable. In practice this usually means choosing the lowest possible wavelength that will provide sufficient signal strength (Fig. 5.12).

5.8 Signal processing for radiation pyrometry

Where a radiation pyrometer is used for measurement of the temperature of a moving object, the signal typically obtained is quite noisy, mainly due to variations in surface conditions along the object.

Signal conditioning will sometimes consist of simple averaging to reduce the influence of local effects on the temperature measurement. (The response time of a typical radiation pyrometer is about 30 ms, and this means that it will follow faithfully a noisy temperature profile, with consequent noise problems for ensuing control circuits. A facility to smooth the signal by averaging over a specified time period allows the user to remove high-frequency noise at the expense of increasing the system response time.)

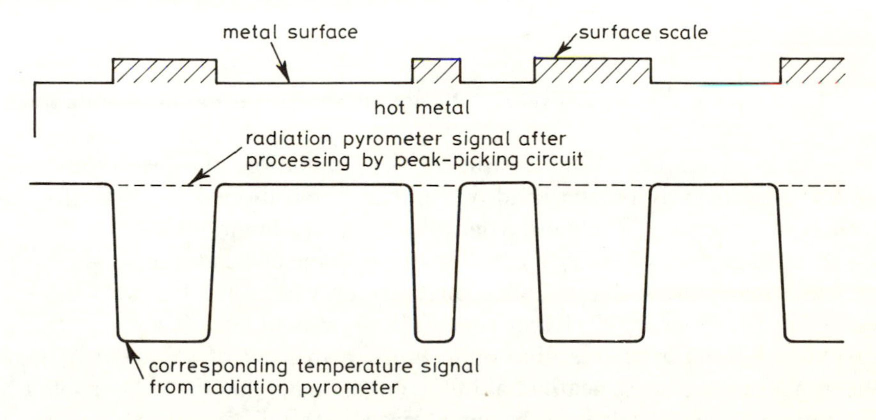

Fig. 5.13 *Peak-picking scheme for measuring surfaces of metal with intermittent lower-temperature surface scale intervening*

In cases, for instance in high-temperature rolling of metals, where it is known that the metal surfaces are partially covered in surface scale, a 'peak-picking' signal conditioner can be used with advantage in an attempt to obtain the measurement attributable to an idealised scale-free object. Such a peak-picking signal conditioner needs to have a fast rise time and a slow decay time chosen to match the dynamics of the particular measurement problem (see Fig. 5.13). A peak-picking circuit is also useful in any application where the line of sight between pyrometer and target might be temporarily obstructed or where a succession of small parts with gaps between is to be monitored as they pass across the target field.

The converse of peak picking, i.e. valley picking, is used in the special application of measuring the temperature of the blades of rotating turbines during operation. This approach is used because the turbine blades are colder than any of the gases or flames that are otherwise in the field of vision of an inserted radiation pyrometer. By operating such a valley-picking pyrometer in stroboscopic synchronism with the turbine rotation, it is possible to obtain the temperatures of individual blades or even of parts of individual blades.

Commercial pyrometer signal-processing units are microprocessor-based with selectable options for signal averaging, peak picking or other variants, together with compensation for surface emissivity and calculation of surface temperature to produce an analysed signal representing true body temperature.

5.9 Built-in calibration for radiation pyrometers

Radiation pyrometers often contain a built in calibration facility. It takes the form of a precisely controlled standard tungsten-filament lamp whose temperature is accurately known. In calibration mode, the radiation sensor is illuminated by the standard lamp, and any discrepancy between the measurement obtained and the standard value is compensated automatically.

5.10 Extension of the capability of radiation pyrometers using fibre-optic links

Fibre-optic techniques allow separation of a radiation pyrometer into two halves: the robust pyrometer head, which can be positioned in physically difficult environments, and the electronic signal processing module.

The arrangement typically comprises the pyrometer head, consisting of an optical system and a silicon radiation detector whose spectral sensitivity is restricted to the band 0·7–1·1 microns. The output of the silicon detector is transmitted along a flexible fibre-optic link constructed of a bundle of glass fibres enclosed in a steel sheath of about 5 mm diameter and up to 3 m in length. The fibre optic-link leads into the signal-processing unit.

The whole system may be considered to be a physically convenient partitioning of a conventional radiation pyrometer into separate optical and signal-processing units, allowing much increased freedom for the applications engineer in siting the sensing head in difficult locations. The upper temperature limit for the optical sensing head and the fibre-optic link is about 200°C, and air cooling may be arranged in particularly difficult environments.

5.11 Commercially available radiation pyrometers

Radiation pyrometers are available to measure temperatures in the range 450–2000°C to an accuracy of better than about 0·5% (expressed as a percentage of the absolute temperature). They can be designed for fixed or hand-held portable use. They are available in combination with signal-processing hardware for averaging, peak picking or valley picking and in combination with 3-term controllers to achieve closed-loop temperature control. Many application-specific devices, e.g. for use in the plastics industry or for measurements of rotating machines, will be found in manufacturers' literature.

5.12 Scanning systems that produce graphical displays of temperature distribution

Infra-red thermography has been used increasingly in clinical practice to detect the temperature-distribution abnormalities that characterise certain illnesses.

The same techniques are now beginning to be applied to process plant for conditioning monitoring and fault diagnosis and to check the efficiency and uniformity of thermal insulation. The apparatus comprises an infra-red scanner to obtain the required temperature-distribution information together with hardware and software that quantise and display the information in rapidly understood coloured isothermal displays.

5.13 Difficulties and advantages associated with radiation pyrometry

5.13.1 Difficulties

(i) The emissivity of the object is usually unknown, and additionally it may vary with time, making accurate conversion of radiation to temperature difficult.

(ii) The object to be measured may not fill the field of view of the device, resulting in too high or too low a measurement, according to the temperature of the extraneous material in the field of view.

(iii) Radiation from outside the nominal field of view may enter the measuring device due to non-perfect cut-off at the boundary and/or due to reflected radiation from extraneous objects.

(iv) Gases, flames and smoke along the line of sight may attenuate or enhance the radiation from the intended target. In either case the result will be an incorrect temperature measurement.

(v) An extraneous object may appear in the line of sight of the device, causing temporary or permanent loss of valid temperature information.

5.13.2 Advantages

Radiation pyrometry offers the possibility to measure the temperature of any object that the pyrometer can be sighted on. A very wide temperature range is covered and very fast response times are possible. Signal-processing techniques allow difficult signals to be 'interpreted'.

5.14 References and further reading

BENEDICT, R. P. (1977): 'Fundamentals of temperature, pressure and flow measurements', (Wiley, New York)

Manufacturers' literature is particularly good in the area of radiation-based temperature sensors. It contains a wealth of technical information not easily obtainable elsewhere.

Chapter 6

New developments in temperature sensors

6.1 Introduction

Nearly every physical phenomenon is, to a greater or lesser extent, temperature dependent. For instance, many electronic, optical, acoustic and magnetic phenomena are highly temperature sensitive. The availability of low-cost small-size integrated circuits allows temperature information to be selectively extracted from signals produced by such phenomena. In this way many 'new' types of temperature sensors may be constructed.

The motivations for developing new types of sensors differ widely and include:

- The wish to produce robust low-cost small-size sensors for inclusion in consumer devices
- The wish to produce extremely accurate calibration standards
- The need for devices that measure mean temperature (rather than spot temperatures) in a spatial region
- The wish to develop robust miniature sensors that can be used to measure temperatures in applications where currently available sensors are not applicable.

6.2 Acoustic thermometry

The velocity of sound in a gas is temperature dependent, and this provides a basis for acoustic measurement of temperature. Such measurement has a number of attractions: it offers a measure of mean temperature along the acoustic path and, by the use of multiple intersecting paths, an estimate of temperature distribution in a region may be obtained.

To infer the temperature of a gas from the velocity of sound, the molecular weight of the gas must be known to a reasonable accuracy and the gas must have a zero mean velocity. Under these conditons, sound pulses are transmitted into the region to be measured and the time of flight of the pulses across a known distance is measured. The method has been used successfully to measure the temperature inside combustion chambers. Where gas velocities are not negligible, it is possible to measure the velocity of sound with and against the gas flow to obtain a true temperature. In the ensuing calculation an estimate is obtained of gas velocity.

6.2.1 Principles

The velocity v of sound in a gas is given by

$$v = \frac{(kT)^{1/2}}{M} \tag{6.1}$$

where k is a constant, T is absolute temperature and M is the molecular weight of the gas.

Eqn. 6.1 shows that temperature in a gas may be inferred from a measurement of the velocity of sound in the gas. To measure the temperature inside a large high-temperature furnace, sound pulses are emitted into the furnace interior from a transmitter and they are received by several acoustic sensors located at different positions. The time-of-flight information allows both mean temperature and temperature distribution to be determined. A number of sources of potential errors exist:

(i) The gas composition may not be known accurately nor may it be homogeneous along the sound paths.
(ii) The sound paths may be curved due to refraction.
(iii) The effect of fast moving gases may modify the basic relation of eqn. 6.1.

Acoustic thermometry is also applied where path lengths are too short to allow accurate time-of-flight measurement. In these applications, phase measurement or detection of change in resonant frequency is used to obtain the necessary acoustic data from which temperature is inferred.

6.2.2 Applications and recent developments

Vetrov (1981) reports an application in which the average air temperature in a buried pipeline has been measured acoustically. A 1000 Hz oscillator drives a loudspeaker to excite the system and a condenser microphone measures the response at the other end of the length of pipeline. A phase meter measures phase shift between the received signal and a reference signal. The phase shift is a function of average temperature, pressure and humidity in the pipeline. Interpretation of the results is claimed to result in an accuracy of 0·01°C in temperature measurement.

Zapka (1982) describes new experimental developments in the simultaneous measurement of velocity and temperature in a flowing fluid. The method uses a laser to generate acoustic pulses of 10^{-9} s duration, and it is the propagation of these pulses that is measured to yield the temperature and flow information. The approach as described can be used only in pure, particle-free fluids.

Meisser (1981) describes a new heat meter that operates by determining the difference in transit time between ultrasonic signals that travel upstream and downstream in a pipe that contains a flowing fluid. Interpretation of the observations yields both temperature and flow information.

Lakshminarayanan (1979) describes a practical ultrasonic device for measurement of gas temperature inside an internal combustion engine. A spark generates the ultrasound and two quartz crystals are used as receivers. Measurement of the time for pulses to travel with and against the gas flow yields both temperature and flow information.

Tasman (1979) describes an ultrasonic thermometer in which the velocity of sound in a thin refractory metal wire is measured and related to temperature. Temperatures up to 2500°C have been monitored for several hundred hours with an estimated uncertainty of ± 30 deg C. The application to measurement of temperature distribution in nuclear reactors is discussed, and it is envisaged that a single sensor suitably configured, may furnish temperature-distribution data from a single ultrasonic measurement.

Gopsalai (1983, 1984) reports work on ultrasonic temperature measurement using a multi-zone sensor. A thin rod, divided by notches into zones, is inserted into the medium to be measured. An ultrasonic wave, applied to the end of the rod, is partially reflected back at each notch, and measurement of the time interval between reflected signals gives a measure of average temperature in the corresponding zone. The application of the technique to coal-gasification environments is discussed.

Satyabala (1981) reports experimental work on the measurement of temperature in a silicon wafer, using acoustic techniques. An externally generated acoustic wave undergoes a temperature-dependent phase shift as it passes through a heated region. The principle is used to determine temperature variations and high sensitivity is reported.

Gujral (1984) shows that an accurate flat and parallel piece of pure metal crystal may be used to measure temperature changes as small as 10^{-6} deg C. The principle is to measure the phase shift of an acoustic wave that is transmitted through the device. The phase shift may be related to temperature-dependent changes in the dimensions of the metal crystal, and hence temperature information may be derived.

Husson (1982) describes a new acoustic technique for the non-invasive measurement of temperature distribution within a solid body. The principle of the method is to determine phase differences between co-linear acoustic signals. Delineation of small temperature differences of a few tenths of a degree Centigrade over spatial differences of 10^{-5} m are reported. A possible application is to the measurement of temperature distribution in small regions of tissue, deep within the body, during treatment by hypothermia.

Rogers (1982) describes experimental work on the use of an immersion probe that yields information on the level, temperature and density of the fluid in which it is immersed. The probe consists of a waveguide which is interrogated alternately by torsional and extensional waves from magnetostriction transducers. The resulting data are processed by computer to obtain level, temperature and flow information.

Kneidel (1982) describes new titanium-ribbon ultrasonic thermometers that have fast response, high durability and good signal/noise ratio.

Satyabala (1981) uses the sing-around technique to obtain high-accuracy measurement of ultrasonic velocity in a liquid. In liquids where velocity is constant, the technique can be used to derive an accurate measurement of liquid temperature.

Capani (1981) describes experimental work in which an acoustic wave is passed through the sample whose temperature is to be measured. Temperature-dependent thermal noise interacts with the acoustic wave as it passes through the sample (a phenomenon known as phonon interaction). After its transit through the sample, the energy content of the acoustic wave is determined. The obtained information is equivalent to a temperature-dependent 'signature' from the sample, and subsequent processing gives a measure of the absolute temperature.

Tam (1983) describes experimental work on the optical measurement of photoacoustic phenomena with possible application to measurement in very difficult environments, such as inside flames and in corrosive liquids. The basic ideas advocated centre on the optical interrogation of ultrasonic phenomena that can be related to process temperature and other process variables.

Moldover (1979) reports experimental work on the use of radial acoustic resonators, in conjunction with velocity measurements, to determine accurately the temperature of dilute gases.

Krylovich (1979) describes theoretical investigations into the possible use of the non-stationary Doppler effect to measure thermal variables.

6.3 Semiconductor-junction temperature sensors

The voltage across the forward-biased junction of a semiconductor is temperature dependent and therefore offers a possible means of temperature measurement. The attraction of temperature sensors based on this principle is their low cost, very small size and their suitability for easy incorporation into integrated circuits.

Both the apparent disadvantages of nonlinearity and device-to-device variability can be completely overcome by suitable design of the circuit into which the semiconductor is incorporated. One approach to this compensatory design is the switching method (there are variants based on the same principle due to Verster (1968). In this method, the collector current of the semiconductor is switched periodically between two fixed values. The ratio of the voltage across the semiconductor junction under the two conditions is then a function only of universal constants and a linear function of absolute temperature. A typical commercially available device operating on the principles described has the specification:

Temperature range: −25 to +105°C
Output: linear current source at 1 μA/°C
Cost: under $1·00 for large quantities
Size: 5 × 5 × 3 mm

6.4 Photoluminiscent fibre-optic temperature sensor

In the photoluminescent fibre-optic sensor, light with a known spectral distribution is transmitted from a source along a fibre-optic link to a photoluminescent

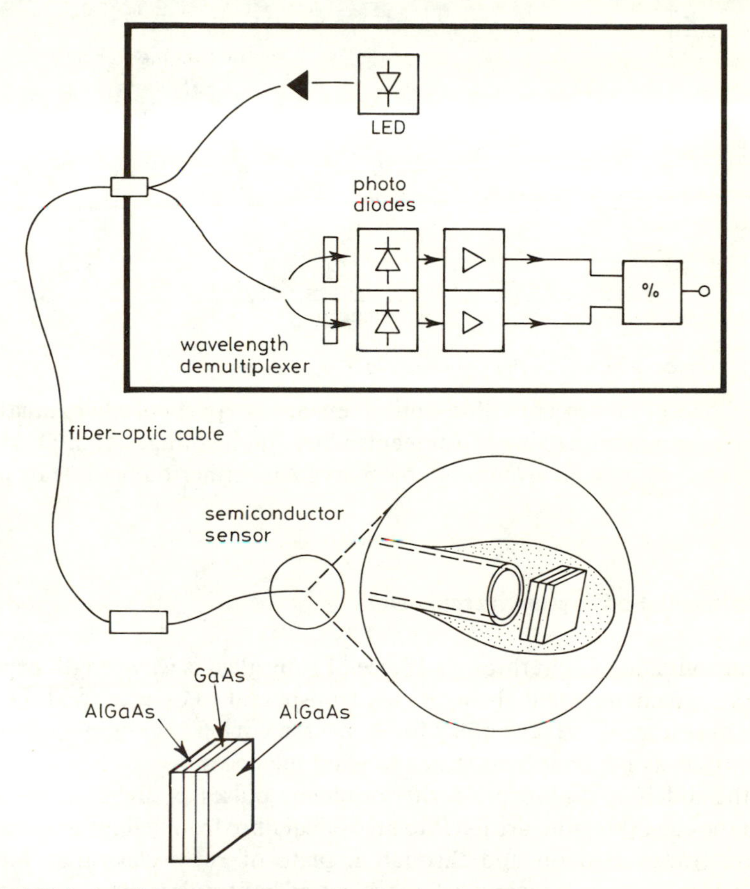

Fig. 6.1 *ASEA photoluminescent temperature sensor*

crystal that is the temperature sensor. The crystal retransmits light back along the fibre-optic link and the frequency of the retransmitted light depends on the temperature of the crystal.

In the ASEA realisation of the device the gallium-arsenide photoluminescent sensor crystal is only 180 microns square and is housed in a 0·55 mm-diameter probe (Figs. 6.1 and 6.2). The fibre-optic link has a quartz core of 100 microns diameter. It can be up to 500 m long. The overall measuring system operates over the range 0–200°C, with resolution 0·1 deg C and an accuracy of ± 1 deg C.

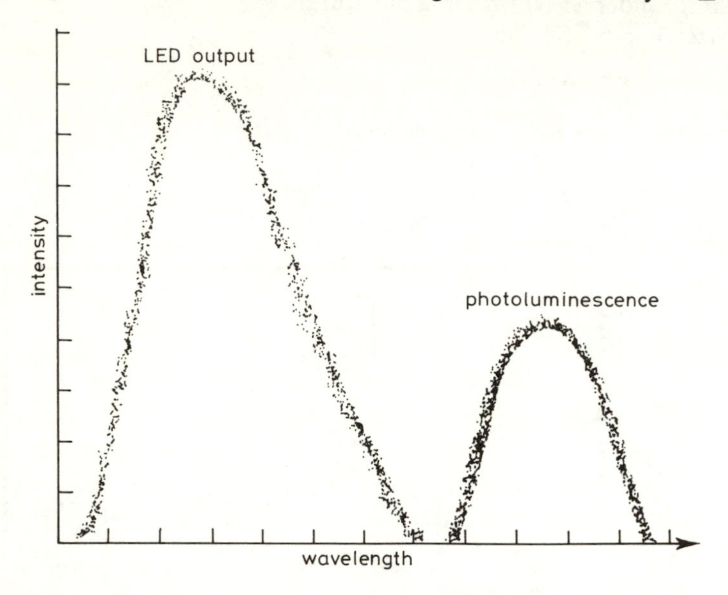

Fig. 6.2 *Principle on which the device of Fig. 6.1 operates*

The photoluminescent fibre-optic sensor is particularly valuable for measuring temperatures in environments where high voltages, radio frequencies or microwaves make conventional measurements either dangerous or liable to errors.

6.5 Differential-absorption thermometer

The transmission of light through a piece of ruby glass is wavelength dependent. The material changes from being highly transparent to opaque, and the change occurs over a small wavelength. Moreover, the cut-off region moves to longer wavelengths as the temperature of the glass increases.

In the differential-absorption thermometer, pulses of light, of wavelength within the cut-off region, are fed into an optical fibre from a light-emitting diode and are transmitted on and through a plate of ruby glass that forms the temperature sensor. A second reference set of light pulses, of a wavelength to which the ruby glass is always transparent, is multiplexed with the first light signal. Comparison of the two signals yields information on the temperature of the ruby-glass sensor. Any unknown or varying attenuation in the optical-fibre lead is automatically compensated for by the reference signal.

6.6 Determination of process temperature based on measurement of nitrogen oxides in combustion gases

Considerable interest centres on nitrogen oxides as pollutants, and correlations have been established between combustion temperatures and the formation of nitrogen oxides.

It also follows that in processes where temperature is particularly difficult to measure, the nitrogen-oxide content of the waste gases might provide an indirect measure of process temperature. This idea has been investigated in trials reported by Ketton Cement and Hartmann & Brown. In the trials, an ultraviolet analyser was installed to measure nitrogen oxide from a kiln, and successful results were reported indicating that kiln temperature can indeed be inferred from analysis of nitrogen oxides in the waste gases ('Control in action' 1983).

6.7 Noise thermometry

This uses the principle that thermally generated noise in a resistor generates an EMF that can be measured and related to temperature. Noise thermometry yields absolute temperatures on the thermodynamic scale, and hence a sufficiently accurate device could form a primary standard. Accuracies of 0·1% in temperature measurements up to 1000°C are reported.

6.8 Mossbauer thermometry

This uses the principle that the width of lines in the spectrum of gamma emission is temperature dependent.

6.9 Magnetic thermometry

This is well established as a tool for the measurement of temperatures close to absolute zero, and is also under development for the measurement of typical industrial temperatures. One such device under development uses the temperature-dependent magnetisation of iron whiskers, and in particular uses the second-harmonic response to yield a temperature reading.

Safrata (1980) describes a specialist experimental magnetic thermometer for measurement of very low temperatures. Rusby (1980) describes experimental work on magnetic measurement of temperatures between 0·4 and 3·1 deg K. Arrott (1981) describes the possible application of magnetic thermometers to the measurement of high temperatures. The principle is again to make use of the second harmonic response of magnetic whiskers that form the sensor elements.

6.10 Nuclear quadrupole resonance (NQR)

The resonant frequency within certain crystalline lattices is temperature sensitive and the relationship is very stable, making it potentially usable as a temperature calibration standard.

One available NQR precision thermometer uses the resonant frequency of a potassium-chlorate crystal in this way. The resonant frequency changes by about 5 kHz per degree K at room temperature, and the device offers a primary temperature standard over the range 90–398 deg K with $\pm 1{\cdot}3 \times 10^{-3}$ deg K reproducibility.

There is also commercially available a very high resolution (0·0001 deg C) sensor, operating over the range −80 to +250°C using change of resonant frequency with temperature of a precisely cut quartz crystal as the sensing means. Because frequency can be measured very precisely, the approach offers the accuracy and repeatability appropriate to a standardising laboratory.

6.11 Fluorescence thermometry

The rate of decay of fluorescence is temperature-dependent. Sensors exploiting this principle are under development, with fibre optics providing the means whereby the rate of decay information can be brought out from the sensor.

6.12 Review papers

Coates (1979) reviews modern methods for the determination of thermodynamic temperature. The methods reviewed include:

(i) Gas thermometry
(ii) Acoustic thermometry
(iii) Noise thermometry
(iv) Photoelectric pyrometry
(v) Radiation calorimetry.

Tasman (1979) discusses three unconventional instruments for the measurement of temperature greater than 700°C using the principles of:

(i) Mossbauer thermometry; discussed in Section 6.8
(ii) Noise thermometry; discussed in Section 6.7
(iii) Ultrasonic thermometry; this is discussed in relation to its application in tracking the rapid variations of gas temperature in the cylinders of an internal combustion engine.

6.13 References and further reading

ANTAL, A. A. (1980): 'Temperature measurements in moving coils', *Kep- & Hangtech.*, **26,** pp. 54–56

ARROTT, A. S., and HEINRICH, B. (1981): 'Application of magnetization measurements in iron to high temperature thermometry', *J. Appl. Phys.*, **52,** pp. 2113–2115

BLALOCK, T. V., and SHEPHARD, R. L. (1982): 'A decade of progress in high temperature Johnson noise thermometry' *in* SCHOOLEY, J. F. (Ed.): 'Temperature: its measurement and control in science and industry'. Vol. 5 (American Institute of Physics)

BRIXY, H., HECKER, R., RITTINGHAUS, K. F., and HOWENER, H. (1982): 'Application of noise thermometry in industry under plant conditions' *in* SCHOOLEY, J. F. (Ed.): 'Temperature: its measurement and control in science and industry'. Vol. 5 (American Institute of Physics)

CAPANI, P. M. (1981): 'Temperature measurements in solids by phonon interaction', *IBM Tech. Disclosure Bull.*, **23,** p. 4480

COATES, P. B. (1979): 'The techniques and relevance of thermodynamic temperature measurement', *High Temp. High Pressures,* pp. 119–134

COLCLOUGH, A. R. (1982): 'Primary acoustic thermometry' (Institute of Physics)

CONTROL IN ACTION (1983), Process Engineering, p. 19, January

DECRETON, M. C. (1982):'High temperature noise thermometry for industrial applications' *in* SCHOOLEY, J. F. (Ed.): 'Temperature: its measurement and control in science and industry'. Vol. 5 (American Institute of Physics)

DADD, M. W. (1983): 'Acoustic thermometry in gases using pulse techniques', *High Temperature Technology,* **1,** pp. 333–342

HUSSON, D., BENNETT, S. D., and KINO G. S. (1982): 'Remote temperature measurement using an acoustic probe', *Appl. Phys. Lett.*, **41,** pp. 915–917

GOPALSAMI, N., and RAPTIS, A. C. (1983): 'Simultaneous measurement of ultrasonic velocity and attenuation in thin rods with application to temperature profiling'. Ultrasonics Symposium Proceedings, Atlanta, GA, USA, Vol. 2, pp. 856–860

GOPALSAMI, N., and RAPTIS, A. C. (1984): 'Acoustic velocity and attenuation measurements in thin rods with application to temperature profiling in coal gasification systems', *IEE Trans.*, **SU-31,** pp. 32-39

GUJRAL, R. (1984): 'Acoustic thermometry', *J. Pure & Appl. Ultrason. (India),* **6,** pp. 21–24

KAPLUN, I. M., LEVEDEV, Yu. P., MARKOV, F. V., and UKHLINOV, G. A. (1983): 'Light-guide pyrometer for monitoring high temperatures', *Izmer. Tekh (USSR),* **26,** pp. 38–39

KNEIDEL, K. E. (1982): 'Advances in multizone ultrasonic thermometry used to detect critical heat flux', *IEEE Trans.*, **SU-29,** pp. 152–156

KRYLOVICH, V. I (1979): 'Nonstationary Doppler effect and frequency-phase methods of investigation and control', *Inzh. Fiz. Zh (USSR),* **36,** pp. 487–492

KYUMA, K., TAI, S., SAWADA, T., and NUNOSHITA, M. (1982): 'Fibre optic instrument for temperature measurement', *IEEE J. Quantum Electron.*, **QE-18,** pp. 676–679

LAKSHMINARAYANAN, P. A., JANAKIRAMAN, P. A., GAJENDRA-BABU, M. K., and MURTHY, B. S. (1979): 'Measurement of pulsating temperature and velocity in an internal combustion engine using an ultrasonic flowmeter', *J. Phys. E.*, **12,** pp. 1053–1058

MANLEY, T. (1979): 'Nuclear magnetic susceptibility thermometry at ultralow temperatures'. Paper GO 14, American Physical Society General Meeting, Chicago, Ill., USA

MARTIN, K.R. *in* JERRARD, H. G. (1982): 'The use of fibre optics in engine pyrometry'. Electro-Optics/Laser International '82 Conference Proceedings, Brighton, England (Butterworth) pp. 18–32

MEISSER, C. (1981): 'A heat meter utilizing static measurement of flow rate', *Landis & Gyr Rev.* **28,** pp. 20–25

MOORE, G. (1984): 'Acoustic thermometry', *Electron. & Power,* **30,** pp. 675–677

MOLDOVER, M. R., WAXMAN, M., and GREENSPAN, M. (1979): 'Spherical acoustic resonators for temperature and thermophysical property measurements'. *High Temp. High Pressures,* **11,** pp. 75–86

OHNO, J., NAKAMURA, M., MIYABE, Y., KAWASAKI, A., and KANOSHIMA, Y. (1982): 'Differential thermometer for measuring hot gas temperature' *in* SCHOOLEY, J. F., (Ed.): 'Temperature: its measurement and control in science and industry'. Vol. 5 (American Institute of Physics)

OHTE, A., and OWAOKE, H. (1982): 'A new nuclear quadrupole resonance standard thermometer' *in* SCHOOLEY, J. F. (Ed.): 'Temperature: its measurement and control in science and industry' Vol. 5. (American Institute of Physics)

ROGERS, S. C., and MILLER, G. N. (1982): 'Ultrasonic level, temperature and density sensor', *IEEE Trans.,* **NS 29,** pp. 665–668

RUEHLE, R. A. (1975): 'Solid state temperature transducer outperforms previous transducers', *Electronics,* **48,** pp. 127–130

RUSBY, R. L., and SWENSON, C. A. (1982): 'A new determination of the helium vapour pressure scales using a CMN magnetic thermometer and the NPL-75 gas thermometer scale', *Metrologia,* **16,** pp. 73–87

SAFRATA, R. S., KOLAC, M., MATAS, J., ODEHNAL, M., and SVEC, K. (1980): 'Deuterized cerium lanthanum magnesium nitrate as a magnetic thermometer', *J. Low Temp. Phys.,* **41,** pp. 405–407

SANO, T., YAMADA, T., ANDO, S., and WAATANABE, K. (1982): 'Temperature measuring method by using the eddy current technique', *in* SCHOOLEY, J. F. (Ed.): 'Temperature: its measurement and control in science and industry'. Vol. 5 (American Institute of Physics

SATYABALA, S. P., SAMPATH-KUMAR, J., MADHUSUDAN RAO, B., and MALLIKARJUN RAO, S. P. (1981): 'Ultrasonic temperature controller using the sing-around technique'. *Acustica* **49,** pp. 163–164

SEKI, K., and MURAKAMI, K. (1980): 'Indicating thermometers using a temperature-sensitive magnetic core'. International Magnetics Conference (INTERMAG) Boston, Mass. USA, Paper 5–9

STEARNS, R., KHURI-YAKUB, B. T., and KINO, G. S. (1983): 'Phase-modulated photoacoustics'. Ultrasonics Symposium Proceedings, Atlanta, Ga, USA, Vol. 2, pp. 649–654

SYDENHAM, P. H. (ed.) (1983): 'Handbook of measurement science', Vol. 2 (Wiley, Chichester, UK) p. 1162

TASMAN, H. A., and RICHTER, J. (1979): 'Unconventional methods for measuring high temperatures', *High Temp. High Pressures,* **11,** pp. 87–101

TASMAN, H. A. *in* DEKLERK, J., and MCAVOY, B. R. (Eds.) (1979): 'Nuclear applications of ultrasonic thermometry'. IEEE Ultrasonics Symposium Proceedings, New Orleans, LA, USA, pp. 950, 380–383

TAM, A. C., ZAPKA, W., COUFAL, H. and SULLIVAN, B. (1983): 'Optical probing of photoacoustic propagation for noncontact measurement of flows, temperatures and chemical compositions' *in* 'Third International Topical Meeting on Photoacoustic and Photothermal Spectroscopy, Paris, *J. Phys. Colloq. (France),* **44,** pp. 203–207

VERSTER, T. C. (1968: 'p–n junction as an ultralinear calculable thermometer', *Electron. Lett.,* **4,** pp. 127–130

VETROV, V. I., and DUBROV, M. N. (1981): 'An acoustic method for measurement of the average air temperature in a beam guiding line', *Radiotekh & Elektron (USSR),* **26,** pp. 2432–2441

YAGOOBI, J. S., CROWLEY, J. M., and CHATO, J. C. (1983): 'Viscometric temperature measurement in electric or magnetic fields'. ASMR/JSME Thermal Engineering Conference Honolulu, HI

ZAPKA, W., and TAM, A. C. (1982): 'Non contact optoacoustic determination of gas flow velocity and temperature simultaneously', *Appl. Phys. Lett.,* **40,** pp. 1015–1017

Chapter 7

Commonly arising industrial temperature-measurement problems

7.1 Introduction

Many application-dependent problems arise because of the physical configuration of a particular process. Some commonly arising features that need non-routine consideration are:

(i) The surface temperature of a solid (perhaps a solid that is in rapid motion) needs to be measured.
(ii) The average temperature in a region needs to be measured accurately.
(iii) The temperature of a moving gas needs to be measured.
(iv) The internal temperature within an impenetrable massive solid needs to be measured.

7.2 Measurement of surface temperature

The measurement of surface temperature of solids (metallic, non-metallic, rough, smooth, flat, curved, stationary, moving) requires special techniques. Applications arise in the paper, textiles and foil industries.

The direct approach is to use special thermocouples of low thermal mass, matching the shape of the surface to be measured, and to use spring pressure to ensure good thermal contact between thermocouple and surface.

A second approach is to fix a small enclosure that produces near black-body conditions over the surface to be measured and then to use an infra-red sensor to monitor the radiation level and to relate this to surface temperature.

Another approach is to use a device such as the Hartmann & Braun thermo-vortex generator, the 'thermoturbolator' (Fig. 7.1). The thermoturbolator is a non-contact device that uses forced convection from the surface of interest to a thermocouple positioned close to the surface. A fan produces a vortex that transmits the temperature across the narrow clearance gap (1 mm) to the measuring area. The physical size is small and the whole arrangement is thermally insulated so that the cooling effect on the surface is small and the response time is rapid.

Normal radiation-pyrometry techniques measure surface temperatures, but for low-temperature workpieces of low and uncertain emissivities, special surface-measuring devices and techniques are often called for.

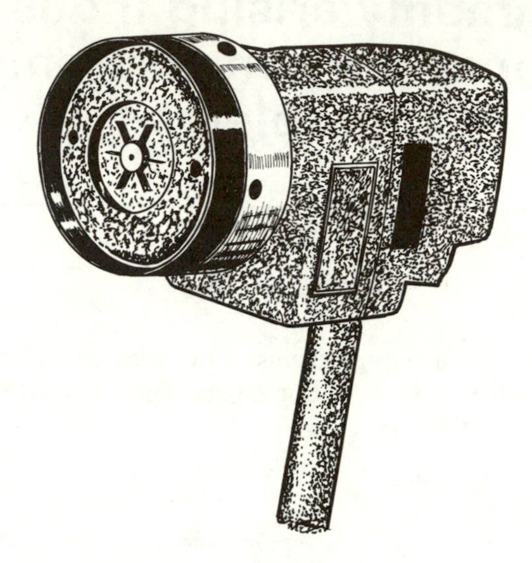

Fig. 7.1 *Thermo-vortex generator (Hartmann & Braun)*

7.3 Determination of the average temperature in a spatial region

Certain applications require the average temperature in a region to be measured, but most sensors give measurements relating to a point or small region in space.

Many large oil-storage tanks are in use throughout the world and they have very accurate level measurement operating to ± 1 mm accuracy. In a typical tank of depth 30 m, there are temperature gradients caused by external climatic conditions or by the addition of new contents. Accurate accounting or dispensing by weight requires the average temperature of the contents to be known. For such applications, long, vertically mounted resistance thermometers may be used. Alternatively, multi-path acoustic measurement may be arranged. Pritchard (1984) describes the application of resistance thermometers to measurement of the average temperature in bulk-storage tanks.

7.4 Measurement of the temperature of a moving gas

An ideal gas-temperature measuring sensor is defined as one that brings a moving gas completely to rest locally and has no heat transfer to its surroundings. The gas temperature indicated by such an ideal sensor is the so-called *total temperature* that would be indicated by a moving probe immersed in and travelling with the gas plus an additional term representing the kinetic energy of the moving gas.

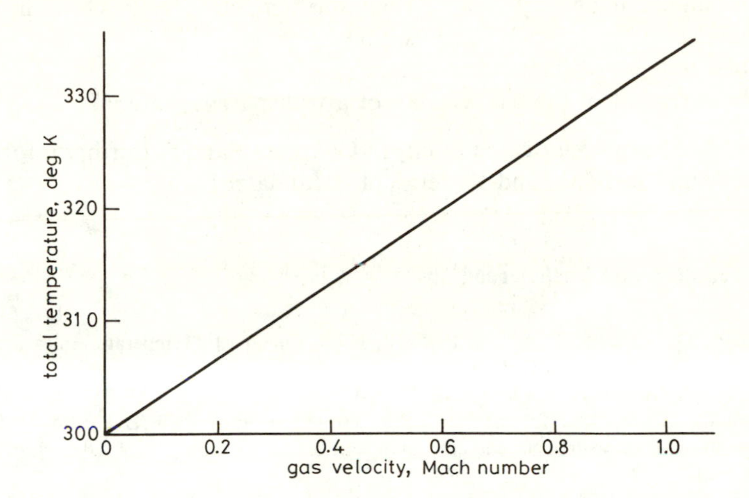

Fig. 7.2 *Total-temperature/Mach-number for a monatomic gas at a static temperature of 300°K*

Let T_0 be the temperature of gas when at rest. Let the gas now be set in motion at Mach number M; then the total temperature T_T of the moving gas is

$$T_T = T_0(0{\cdot}5(C_P/C_V) - 1)M + 1) \tag{7.1}$$

Fig. 7.2 shows the relation given by eqn. 7.1 for $T_0 = 300$ deg K. The Figure shows that, at high gas velocities, T_T is significantly greater than T_0.

7.5 Determination of the internal temperature of an impenetrable massive solid

Throughout the primary-metals industries, large blocks of material are heated in preparation for rolling, forging or other hot working. In such cases, the internal temperatures are of vital importance, but they can never be measured directly on a routine basis.

When conditions are standardised the problem is solved by building up a set of rules that have been found to give satisfactory results.

Where conditions are variable, some type of mathematical model must be used. Such a model may be developed and refined with the aid of data from special trials with instrumental workpieces with imbedded temperature sensors (Wick, 1978).

7.6 Other specialist temperature-measurement problems

Other problems that require special consideration are:

- Measurement of very high or very low temperatures
- Measurement of very rapidly changing temperatures (such as inside the cylinders of an internal combustion engine
- Measurement of flame temperature
- Measurement of the temperature of a transparent material.

These topics have been treated briefly in Chapters 5 and 6. Further information may be found in Baker and Ryder, Vol. 2 (undated).

7.7 References and further reading

BAKER, H. D., RYDER, E. A., and BAKER, N. H. (undated): 'Temperature measurement in engineering' Vol. 2 (Omega Press, USA)

OHNO, J., NAKAMURA, M., MIYABE, Y., KAWASAKI, A., and KANOSHIMA, Y. (1982): 'Differential thermometer for measuring hot gas temperature' *in* SCHOOLEY, J. F. (Ed.): 'Temperature: its measurement and control in science and industry' Vol. 5, (American Institute of Physics)

PRITCHARD, J. S. (1984): 'Averaging temperature in bulk storage applications', *Transducer Technology,* Nov., p. 19

WICK, H. J. (1978): 'Estimation of the core temperature in a soaking put with a modified Kalman filter', p. 548–552 (in German)

Chapter 8

Heat sources

8.1 Brief look at the sources available

An attempt to list possible sources of heat produced the following:

(i) Exothermic reaction
(ii) Release of latent heat during change of state
(iii) Release of heat during pressure changes in a gas
(iv) Mechanical friction
(v) Electrical $i^2 R$ heating
(vi) Electrical heating due to alternating magnetisation
(vii) Electrical heating due to alternating electrification
(viii) Electrical heating due to thermoelectric effects
(ix) Thermonuclear heating.

For the purposes of this practically oriented book we move to the following derived set of readily applicable heat sources:

(*a*) Combustion, from (i)
(*b*) Heat pumps, from (ii) and (iii)
(*c*) Electrical resistance heating, from (v)
(*d*) Induction heating, from (v) and (vi)
(*e*) Microwave heating, from (vii)
(*f*) Electric arc heating, from (v)
(*g*) Infra-red heating, from (v)

Combustion is clearly in practice the most important of the available sources of heat, and therefore we devote considerable space to this topic.

8.2 Combustion fundamentals (Glassman, 1977)

Suppose that a fuel is mixed with an oxydiser and that combustion occurs. If all the heat of reaction is used solely to raise the temperature of the resultant product, then this resultant temperature is called the *adiabatic flame temperature*. The adiabatic flame temperature can be calculated using equations of the form:

$$\text{Fuel A} + \text{Oxydiser B} \longrightarrow \text{Product C} + \text{Heat evolution D} \qquad (8.1)$$

Such equations are most usually expressed in moles of substances.

8.2.1 Use of Tables of heats of formation

Considering elements in their standard states (i.e. at a pressure of one atmosphere and at a fixed temperature: 298·1 K in Table 8.1) to have by definition zero energy of formation, then the energy of formation of any compound may be determined by calorimetry as in the examples in the Table. The use of the Table is illustrated by the examples below:

$$C\,(\text{solid}) + O_2\,(\text{gas}) \longrightarrow CO_2\,(\text{gas}) + 94{\cdot}05\,\text{kcal}$$

$$H_2\,(\text{gas}) + \tfrac{1}{2}O_2\,(\text{gas}) \longrightarrow H_2O\,(\text{gas}) + 57{\cdot}80\,\text{kcal}$$

$$H_2O\,(\text{gas}) \longrightarrow H_2O\,(\text{liquid}) + (68{\cdot}32 - 57{\cdot}80)\,\text{kcal}$$

$$\tfrac{1}{2}N_2\,(\text{gas}) + \tfrac{1}{2}O_2(\text{gas}) \longrightarrow NO\,(\text{gas}) - 21{\cdot}6\,\text{kcal}$$

Table 8.1 *Heats of formation*

Substance	Symbol	Heat of formation at 298·1°C (kcal/mole)
Nitrogen atom in gaseous state	N	112·75
Oxygen atom in gaseous state	O	59·16
Acetylene gas	C_2H_2	54·19
Hydrogen atom in gaseous state	H	52·09
Ethane gas	C_2H_2	12·50
Hydroxyl radical in gaseous state	OH	10·06
Oxygen gas	O_2	0
Nitrogen gas	N_2	0
Hydrogen gas	H_2	0
Solid carbon	C	0
Methane gas	CH_4	−17·89
Carbon-monoxide gas	CO	−26·42
Water vapour	H_2O	−57·80
Liquid water	H_2O	−68·32
Carbon-dioxide gas	CO_2	−94·05

These sample data are from a larger table that may be found in Glassman (1977).

The first two equations show the exothermic combustion reactions for complete combustion of carbon and hydrogen. The third equation represents the latent heat of vaporisation of water. It quantifies the energy loss in waste gases that contain water vapour. The last two equations are the endothermic relations of oxygen dissociation and nitrous oxide formation, respectively. Both reactions can occur at high temperatures such as are encountered in the interior of flames. Both reactions have a significant adverse effect on flame temperature. The nitrous-oxide formation is, in addition, considered to represent a serious pollutant factor, even when it occurs in small quantities. According to Glassman (1977) any temperature above 1800 deg K may produce sufficient nitrous oxide to cause concern in terms of pollution.

Equations representing heat of formation can be combined to show sequential reactions as, for example:

$$C\,(\text{solid}) + \tfrac{1}{2}O_2\,(\text{gas}) \longrightarrow CO\,(\text{gas}) + 26{\cdot}42\,\text{kcal}$$

$$CO\,(\text{gas}) + \tfrac{1}{2}O_2\,(\text{gas}) \longrightarrow CO_2\,(\text{gas}) + (94{\cdot}05 - 26{\cdot}42)\,\text{kcal}$$

Table 8.2 *Examples of peak attainable flame temperatures for some fuel oxidiser mixtures at normal pressures and initial temperatures (298 K)*

Fuel	Oxidiser	Temperature deg K
Acetylene	Oxygen	3410
Acetylene	Air	2600
Carbon Monoxide	Air	2400
Hydrogen	Air	2400
Methane	Air	2210

At higher pressures, higher flame temperatures are attained.

8.2.2 Flame-temperature calculation

Below 1250 deg K, CO_2, H_2O, N_2 and O_2 are stable and flame temperatures may be calculated from data in Table 8.1 in conjunction with a knowledge of the specific heat of the product gases.

At higher temperatures dissociation begins and in a C–H–O system any of the following (endothermic) dissociation reactions might be encountered:

$$CO_2 \longrightarrow CO + \tfrac{1}{2}O_2$$

$$CO_2 + H_2 \longrightarrow CO + H_2O$$

$$H_2O \longrightarrow H_2 + \tfrac{1}{2}O_2$$

$$H_2O \longrightarrow H + OH$$

$$H_2O \longrightarrow \tfrac{1}{2}H_2 + OH$$

$$H_2 \longrightarrow 2H$$

$$O_2 \longrightarrow 2O_2$$

Thus the exact composition of the product of combustion is difficult to calculate precisely.

A graph of flame temperature against fuel/air ratio has the form shown in Fig. 8.1. Certain fuel/oxidiser mixtures have their maximum temperature peak on the rich side of stoichiometric because of specific-heat effects.

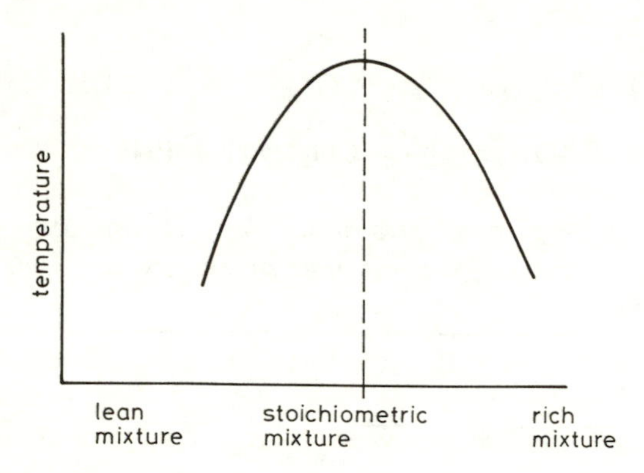

Fig. 8.1 *Variation of temperature with air/fuel ratio*

8.3 Premixed and diffusion flames

Flames can be categorised roughly into two types

(*a*) *Premixed*: Here the fuel and the oxidiser are well mixed before combustion. The combustion is then reaction-rate limited (a Bunsen flame is an example of a premixed flame).

(*b*) *Diffusion*: In a diffusion flame, the fuel and oxidiser are admitted in separate streams into the combustion chamber. The combustion is usually regarded as diffusion-limited. Most industrial flames are diffusion flames.

8.4 How much air is required for complete combustion?

Unnecessary excess air in a combustion process is a common major cause of inefficiency. The excess air is simply heated to the furnace temperature, only to leave carrying sensible heat to waste.

An initial view therefore is that *no* excess air should be present in a combustion process. However, when we consider that combustion necessarily takes place in a spatial region where conditions cannot be perfectly homogeneous, it becomes clear that, if we inject exactly sufficient air to burn a given incoming quantity of fuel, we shall expect to have, because of inhomogeneity in the combustion region, a slight excess of air in some parts of the region and a slight deficiency in other parts. Thus, for instance, oxygen and CO will co-exist, and measurement of their concentrations would be expected to yield information on the degree of inhomogeneity of the combustion process. Three points to note therefore are the following:

(i) If complete combustion should be required, some quantity of excess air will be needed to allow for the inhomogeneity of combustion.

For the ideal combustion of coal, an excess oxygen content of about 8% may be necessary. For oil, with its improved mixing properties, about 5% excess oxygen is needed, while in gas firing an excess oxygen content of less than 3% will usually be necessary if optimum combustion conditions are to be obtained.

(ii) The degree of inhomogeneity in a particular combustion process will depend on: type of fuel, burner design, temperature and degree of turbulence in the combustion chamber.
(iii) Measurements of oxygen and CO in the gases of combustion will yield information on the degree of inhomogeneity of combustion. If the concentration of CO in the combustion gases is high, it could be argued that further excess air is needed.

Pursuing idea (iii) further, it is possible to imagine controlling a combustion process such that the CO in the waste gases was always kept to a particular value. Such a control system would attempt to maintain a particular degree of completeness of combustion, regardless of (for instance) widely differing furnace loading conditions.

After describing some available methods for measuring concentrations of the products of combustion in waste gases, we outline how the measurements are incorporated into automatic loops to control air/fuel ratio.

8.5 Measurement of oxygen in combustion gases using zirconia cells

Early combustion monitoring relied on measuring carbon dioxide in the waste gases, but the method suffers from the disadvantage that the relation between CO_2 measurement and combustion efficiency is dependent on the composition of the fuel. (For instance, if pure carbon is burnt in air containing 21% oxygen, a maximum CO_2 content, of 21% in the waste gases is possible., For a hydrocarbon fuel, the maximum possible CO_2 is clearly lower.)

The measurement of oxygen in the waste gases gives a more direct indication of combustion efficiency. Such a measurement may be made continuously inside a furnace, without sampling, using the zirconia cell. The zirconia cell produces a potential e according to the Nernst equation

$$e = kT \ln (x_1/x_2) \tag{8.2}$$

where x_1 = oxygen concentration in a reference gas
x_2 = oxygen concentration in the furnace
T = absolute temperature
k = constant

In industrial applications, air acts as the reference gas and the system needs no calibration. The principle of the cell is illustrated in Fig. 8.2. The reference gas and the gas to be measured reach the cell through porous platinum electrodes and the sensor EMF is then generated by electrolytic action in the cell. The temperature value required for insertion in the Nernst equation is provided by

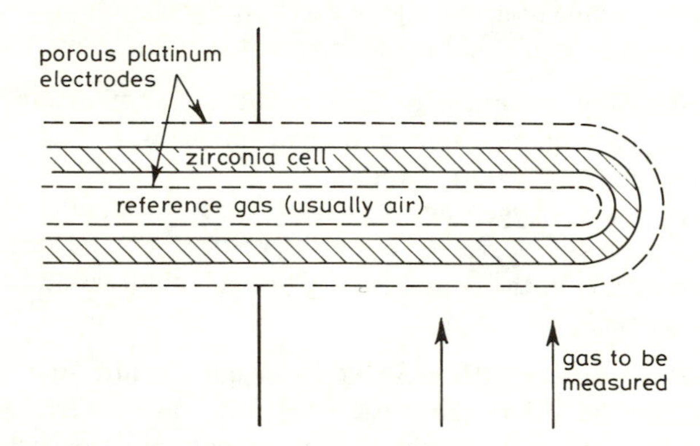

Fig. 8.2 *Zirconia probe for continuous measurement of oxygen in situ*

a thermocouple inserted in the probe housing. A detailed description of the zirconia cell and its application in glass furnaces has been given by Chamberlain (1984).

Where the zirconia cell is to be used in a feedback control system to adjust air/fuel ratio, the siting of the sensor will be particularly important. A location must be chosen that is representative of the combustion conditions, with particular care taken to ensure that the sensor is not affected by local ingress of air. Care must be taken in the control system to produce a safe outcome when the sensor fails. The simplest technique is to limit the changes that the control system is allowed to make to the combustion conditions.

Specifications of zirconia cells vary, but in general they can only operate successfully at temperatures above about 600°C. Their time constants are temperature-dependent but a time constant of 1 s is typical.

8.6 Measurement of carbon monoxide

Gas chromatography is too slow acting (it has a time constant of several minutes) to be useful for on-line control of combustion.

Infra-red absorption techniques, as used for instance in the Hartmann & Brown URAS 2 instrument, give a fast response with a time constant of less than 1 s. Carbon monoxide absorbs infra-red radiation in a particular waveband whereas nitrogen does not absorb at all in this waveband. The principle of the instrument (see Fig. 8.3) is to pass radiation through the sample gas and simultaneously through an identical width of nitrogen gas used as a reference. The two radiation streams then fall onto the two halves of a receiver cell, both filled with pure carbon monoxide and separated by a membrane. The halves of the receiver cell experience a temperature and pressure rise dependent on the intensity of incident radiation. The difference between the pressure in the two

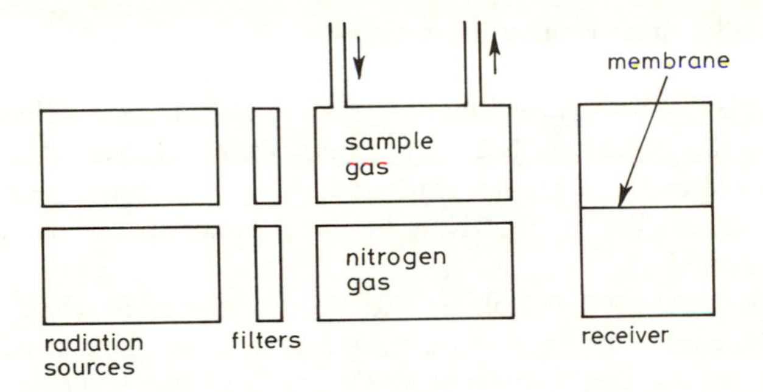

Fig. 8.3 *Principle of operation of carbon monoxide sensor*
Displacement of the membrane is measured capacitively and is calibrated in terms of CO in the sample gas

halves of the receiver is thus a measure of the absorption, and hence of the carbon monoxide in the sample gas. The membrane is flexible and its displacement may be measured by a capacitive transducer calibrated to yield a direct reading of carbon-monoxide content.

8.7 Measurement of sulphur dioxide in flue gases

Sulphur dioxide may be measured *in situ* (i.e. without sampling) in flue gases using a solid-state electrolyte sensor such as the Westinghouse model 260 device. The sensor works in conjunction with an integral oxygen analyser, and in fact the determination of sulphur dioxide involves subtraction of oxygen content from an $SO_2 + O_2$ total. For pollution monitoring, ranges down to 10 p.p.m. of SO_2 may be specified.

8.8 Measurement of nitrogen oxides

Nitrogen oxides have been shown to be important in air pollution. They can be measured by ultra-violet techniques or by mass spectrometer.

8.9 Measurement of all the constituent gases of combustion using a mass spectrometer

High value metallurgical and chemical processes can justify the incorporation of a mass spectrometer into the waste-gas stream, and this gives a rapid, complete analysis that can form an accurate basis for automatic control actions.

8.10 Air pollution from combustion processes

Solid-particle emissions and sulphur oxides are the old established pollutants from combustion processes. Nitrogen oxides were only recognised as major pollutants at a later date when it was realised that photochemical processes used nitrogen oxides as the raw materials in the production of harmful complex organic compounds.

Solid carbon particles in a flame radiate more strongly than the gaseous regions of the flame, and help to increase heat transfer from the flame. Since obtaining maximum heat transfer from flames is a basic requirement in many industrial processes, designers aim to produce particle-bearing flames. As would be expected, solid carbon particles tend to form in the combustion of an over-rich mixture, but they may also be found even when excess oxygen is present. In fact, the mechanism of formation of solid carbon particles is complex, and one of the dominant factors is the molecular structure of the fuel used.

Special burners aimed at producing low quantities of nitrogen oxides are available. Basically they operate by producing a larger combustion region with a lower temperature compared with a normal burner. Pressurised fluidised beds are claimed to remove 80% of sulphur dioxide and to produce less nitrogen oxides than do low-nitrogen-oxide burners.

8.11 Control of combustion processes

Large-scale combustion processes are becoming subject to even more demanding requirements in respect of fuel efficiency and air pollution. Many are being required to burn variable low-quality fuels while minimising emissions of smoke and acid-producing products.

Large coal-burning boilers usually burn pulverised coal directly fed from a mill, since pulverised fuel cannot be stored because of the risk of explosion. Thus the fuel flow to the process is controlled directly by the mill. Where the coal hardness (for instance) is varying significantly, it is difficult to obtain precise control of the fuel feed. This point is discussed again in Section 8.14.2.

The optimum air/fuel ratio for a particular combustion process is not constant but rather is a function of load. This is because burner performance, reaction rates and flow and pressure patterns in the combustion zone are all functions of load. Ideally, therefore, the desired air/fuel ratio needs to be changed as load (or any other influencing variable) changes. Such changes in desired air/fuel ratio may be brought about in response to signals from gas analysers in the waste-gas stream. A more detailed discussion of how this may be achieved is given in the following Section.

Williams (1983) has reported work on large boilers in which an optimal air/fuel ratio is calculated and implemented on-line, based on measurements of smoke, dust, carbon-monoxide and oxygen in the waste gases.

8.12 Quantitative basis for controlling air/fuel ratio to maintain optimum combustion efficiency at varying loads

If, in a particular combustion process, the fuel flow is maintained constant at a high rate while the air flow is moved to give fuel-rich or air-rich conditions, it is found that the losses due to unburned fuel and the losses due to excess air carrying away sensible heat have the forms shown in Fig. 8.4. As expected from previous discussions, optimum combustion (minimum losses) is obtained with some proportion of excess air.

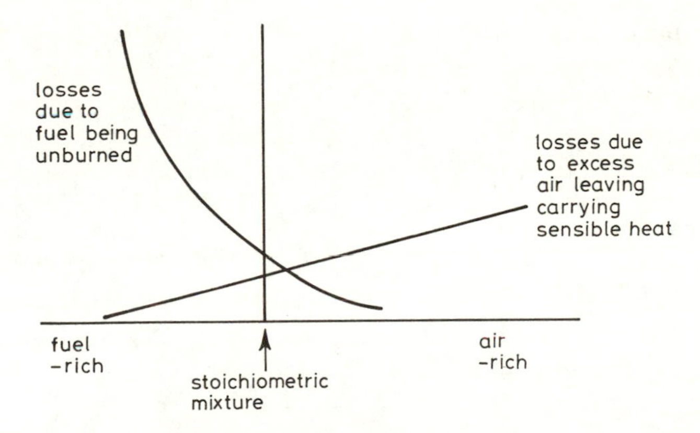

Fig. 8.4 *Heat losses as a function of air/fuel ratio*

Suppose, next, that the above tests are repeated but with a much lower fuel flow. It is found that the (normalised) losses now have the form shown in Fig. 8.5.

The significant difference between the two Figures is that the minimum loss point has moved to the right for the low-fuel-flow (i.e. for the low-heating-load) situation. The reasons for this have been suggested earlier, and ideally a combustion control system should somehow attempt to obtain minimum losses despite changes in loading. A combustion control system that maintained CO content in the waste gases constant at a pre-determined value would appear to offer a good approximation to what is required.

To understand the basis for this statement, assume that the losses due to excess air have the form $J_1 = ax + b$ and that the losses due to unburned fuel have the form $J_2 = ce^{-(x+d)/f}$, where the functions have been chosen to be of the right form to agree with the curves of Figs. 8.4 and 8.5. a, b, c and f are constants whereas d changes to give the load-dependent shift in the unburned fuel curves of Fig. 8.5.

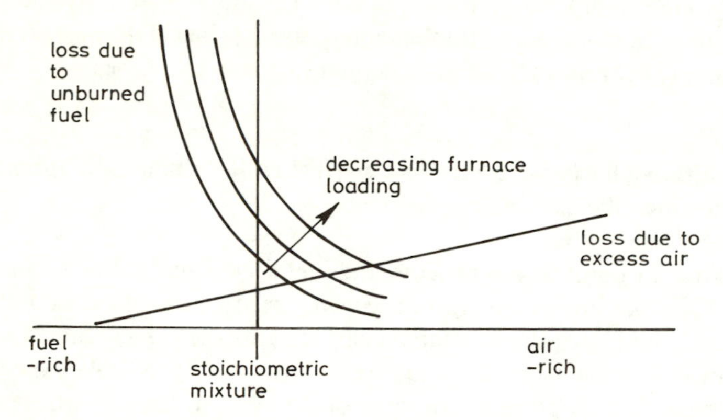

Fig. 8.5 *Heat losses as a function of air/fuel ratio for different values of furnace loading*

Then total loss

$$J = J_1 + J_2$$

$$\frac{\partial J}{\partial x} = a + \left(-\frac{c}{f}\right) e^{-(x+d)/f}$$

For minimum loss

$$\frac{c}{f} e^{-(x+d)/f} = a$$

$$e^{-(x+d)/f} = \frac{af}{c}$$

$$e^{(x+d)/f} = \frac{c}{af}$$

$$\frac{x + d}{f} = \ln\left(\frac{c}{af}\right)$$

and if we denote the x that minimizes J by $\hat{x}$ then

$$\hat{x} = f \ln\left(\frac{c}{af}\right) - d$$

If we substitute the expression for $\hat{x}$, then we can determine that value of J_2 (say $\hat{J}_2$) that corresponds to minimum total loss as

$$\hat{J}_2 = ce^{-(x+d)/f} = ce^{-(\ln c/af - d/f + d/f)}$$

$$= ce^{-\ln c/af} = \frac{c}{c/af} = af$$

Thus we obtain the important result: *the value of J_2 that corresponds to minimum total losses is independent of x, i.e. of fuel flow.*

An application described in Williams (1983) makes use of this principle to attempt to automatically track the point of minimum loss, despite changes in boiler loading. In the application, it has been found that, if the carbon monoxide in the waste gases can be kept between 50 and 300 p.p.m., then the losses will automatically be kept near their minimum possible values. Thus the carbon-monoxide measurement may be used to determine a desired value for excess oxygen in the waste gases.

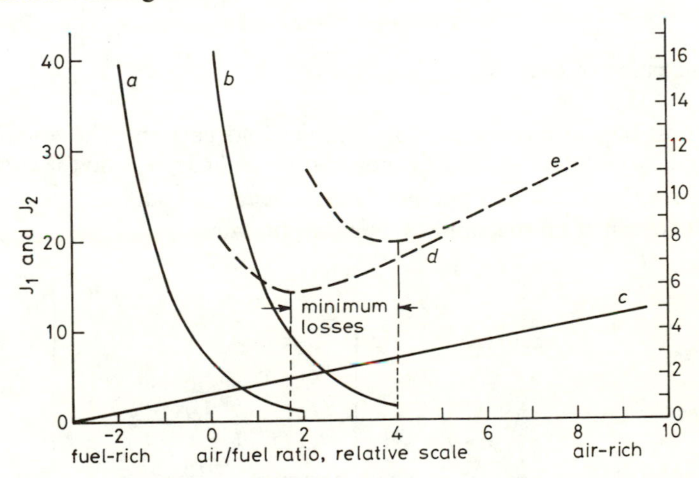

Fig. 8.6 *Total (normalised) heat losses as a function of air/fuel ratio and for two different furnace loads*
(*a*) J_2 for high load conditions
(*b*) J_2 for low load conditions
(*c*) J_1
(*d*) Total losses on high load
(*e*) Total losses on low load

A fuel/air-ratio trim-control system can then be used to maintain the pre-determined oxygen content, thus allowing automatic tracking of the load-dependent minimum-loss point on the curves of Fig. 8.6. A system putting these ideas into practice was installed by Bowater–Scott (UK) in 1981 using Land equipment.

8.13 Maintaining correct air/fuel ratio during transients

In the simple control of a combustion process, a controller manipulates air and fuel valves to achieve a desired temperature and a specified air/fuel ratio. We assume that, in the steady state, the control objectives can be met perfectly, for a wide range of different desired temperatures.

Consider now the transient situation when a step increase in desired temperature occurs (or, equivalently, when a sudden additional heating load is imposed on the system). The two valves controlling air and fuel, respectively, are unlikely to have similar dynamic responses. In fact, because of the smaller physical size of the fuel pipe compared with the air pipe, the fuel valve is likely to respond faster than the air valve, with consequent, possibly severe, lack of combustion air until the new steady state is reached. (During the transient that follows a decrease in desired temperature, the opposite effect will occur, with excess air being delivered.)

The simplest solution is to monitor the position of both air and fuel valves, and to control the fuel valve so that its transient behaviour matches that of the air valve.

8.14 Combustion of coal

The combustibles in coal consist of volatiles and carbon. The volatiles burn rapidly, but carbon itself does not vaporise until 4170 deg K and the combustion takes place within the solid material. The efficiency of coal burning depends on the extent to which all the solid carbon can be burnt.

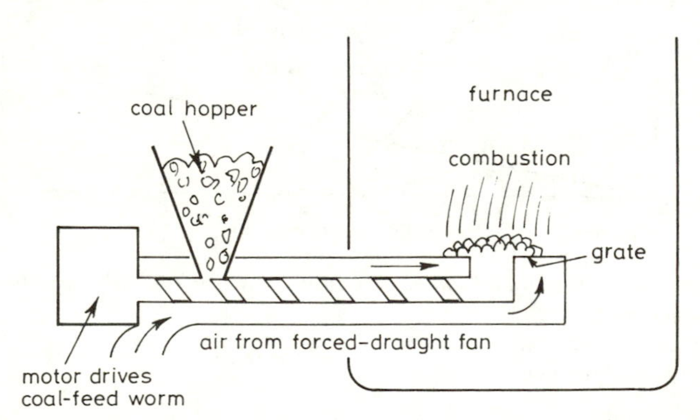

Fig. 8.7 *An underfeed stoker for burning coal*

8.14.1 Underfeed stokers

In an underfeed stoker, coal is fed from a hopper and is driven by a worm along a horizontal tube. This arrangement is totally surrounded and cooled by the combustion air. The coal spills out onto perforated grates where it burns with the aid of the combustion air that is being forced vertically upwards (Fig. 8.7).

Control of the underfeed stoker is obtained by using a motor that simultaneously switches the combustion air between high and low rates and the rotational speed of the coal-feeding worm between high and low rates, using a variable-speed gearbox. The coal/air ratio needs to be set with care not only to obtain optimal combustion as indicated by waste-gas analysis, but also (often) to obtain strongly radiating luminous flames and to establish a satisfactory bed thickness of coal on the grates.

In a high-temperature furnace, the grates rely on the freshly arriving coal and the combustion air to prevent them melting. Therefore it is not allowable as part of the control strategy to shut off the air and fuel feeds to the stoker for more than a very short period. In fact, it is necessary to monitor the operation to ensure that, should the coal feed be interrupted by mechanical breakdown, as may happen when a lump of metal is amongst the coal, then maintenance crews can be alerted before serious damage occurs.

8.14.2 Burning of pulverised coal

All modern coal-fired power stations burn pulverised fuel, produced by on-site coal-grinding mills. Hot air, blown through the mills, entrains the powdered coal and transports it to the burners. Control consists in maintaining both air and coal flow at correct (ratioed) values. Pulverised fuel cannot be stored, since a store of powdered coal is an explosion hazard. Therefore, whenever a coal-fired power station is required to change output, there has to be an immediate change in the output rate of the coal-grinding mill. Since coal hardness varies considerably and unpredictably, it is quite difficult to set up the mill to produce a precisely known mass flow rate of pulverised fuel. It is also quite difficult to measure at all accurately the mass flow from the pulverising mill. The correct setting for the mill (important for efficiency, pollution minimisation and prevention of slagging of the combustion area) therefore has to be undertaken with respect to the measured composition of the waste gases of combustion. Williams (1983) gives detailed information on control strategies for boilers fired by pulverised fuels.

8.15 References and further reading

BINDI, V. D. (1983): 'A predictive and self tuning combustion control system based on in situ gas analysis' *in* 'Advances in Instrumentation', Vol. 38, Pt. 2 Proc. ISA Conference, Houston, Texas, pp. 1151–1161

CHAMBERLAIN, M. (1984): 'Zirconia oxygen sensors for control of combustion in glass melting furnaces', *Glass Technol.*, **25**, (5)

FAITH, W. L., and ATKISSON, A. A. (1972): 'Air pollution' (John Wiley)

GLASSMAN, I. (1977): 'Combustion' (Academic Press)

KYTE, W.S., BETTELHEIM, J., and COOPER, J.R.P. (1983): 'Sulphur oxides control options in the UK electric power generation industry', I. Chem. E. Symposium Series, Vol. 77

RAASK, E. (1985): 'Mineral impurities in coal combustion' (Springer–Verlag)
VDI (GERMANY) (Ed.) (1978): 'Fluidised bed combustion' (Adam Hilger)
WILLIAMS, A. H., and WADDINGTON, J. (1983): 'New automatic control strategies for CEGB boilers', *CEGB Research*, pp. 16–24

Chapter 9

Heat sources: Non-combustion aspects

9.1 Heat pumps

A heat pump is an arrangement in which a work input is used to extract heat energy from a low-temperature source and to deliver as much as possible of that heat energy to a high-temperature sink.

Current commercial heat pumps use a vapour-compression cycle implemented using a compressor, condenser, expansion valve and evaporator. Figs. 9.1–9.3 illustrate, respectively, the principle, the physical layout and the idealised cycle for a heat pump. Control of heat pumps is primarily concerned with

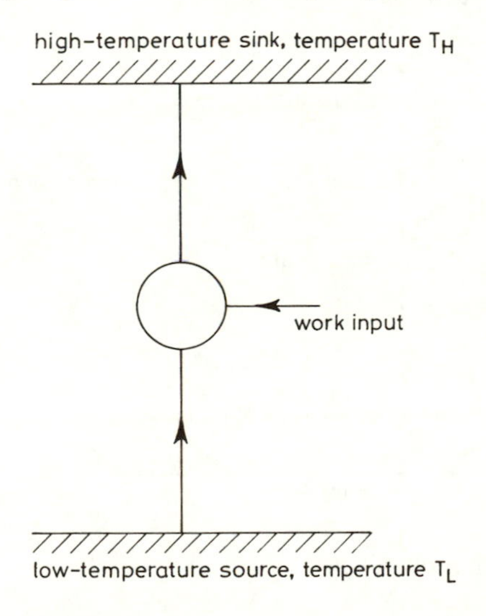

Fig. 9.1 *Principle of a heat pump*

two interlinked problems:

(i) To meet a user's demands under widely varying conditions of loading and ambient temperature

(ii) To maximise the overall efficiency across the wide ranges of conditions implied in (i).

It will be appreciated that the physical components, such as the compressor, have to be sized to meet a peak loading requirement and that it will require a carefully thought-out control strategy to obtain high efficiency from the system under light loading conditions.

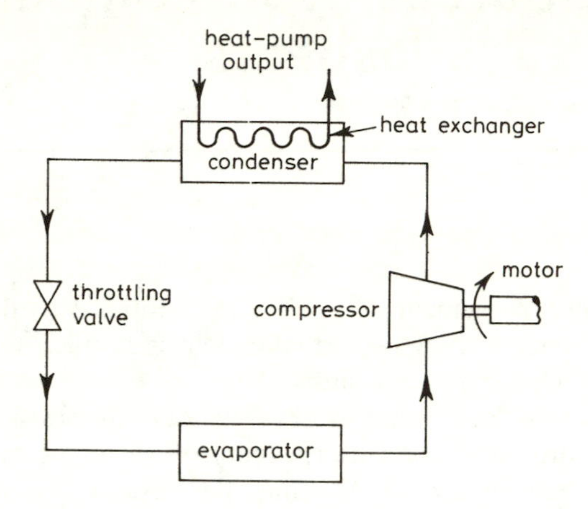

Fig. 9.2 *Schematic of heat pump*

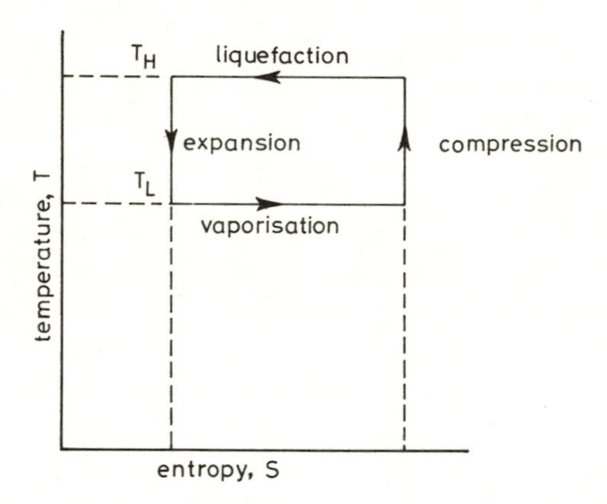

Fig. 9.3 *Idealised cycle for a heat pump*

In designing a control system for any particular heat pump the designer will always need to compromise between aiming for very high efficiency (implying the need for stored models, adaptivity, the maintenance of high efficiency during transients and the choice of a compressor whose speed is continuously variable) and low initial capital cost.

9.2 Direct resistance heating

In direct resistance heating, an electric current is passed, via electrodes or other direct physical contacts, through the material to be heated. The power generated is given by $W = i^2R$, where i is the current and R is the effective resistance between electrodes of the material.

The chief applications of direct resistance heating are:

(i) Rapid heating of metal bars and rods at high current densities, in preparation for hot working
(ii) Water heating and steam raising
(iii) Molten salt baths
(iv) Glass heating.

In the heating of metal bars, the chief problem is to establish reliable low-resistance electrical connections despite irregularities and inconsistencies in the dimensions of the bars. Very rapid heating of bars (from cold to 1200°C in 1 min) is reported from working applications.

Other shapes of metal products, e.g. strips, may also be heated directly, provided that the electrical-connection problem can be overcome. Prototype systems in which the connections are made through liquid metals have been reported.

In direct resistance heating of water, a number of new factors needs to be taken into account. The conductivity of water is highly dependent on the presence of impurities. When a potential difference is applied across electrodes immersed in water, ions as well as electrons are involved in the conduction process. If direct current is used, oxygen and hydrogen will be liberated at the anode and cathode. To avoid this effect, alternating current is used in direct resistance heating of water.

The control of water temperature in a direct-resistance-heating application is achieved by raising or lowering a shield that varies the effective length of the conducting path between electrodes. In contrast, when direct resistance heating is used to raise steam, control of the rate of steam production is achieved simply by altering the water level, since steam is for all practical purposes an insulator.

The conductivity of the water to be heated must be brought to a suitable value as part of the procedure of incoming water treatment. (However, varying the water conductivity is too slow and unwieldy in general to be used in temperature control.) In general, the water treatment for a direct-resistance water heater is more difficult to achieve than for a conventional boiler.

In a normal boiler, feedwater hardness and pH have to be controlled. In the resistance-heating boiler, three water parameters – conductivity, hardness and pH – have to be controlled. The fact that conductivity is dependent on both hardness and pH complicates the feedwater treatment.

9..2.1 Electrical heating elements

ı electrical heating element is typically a band of resistance wire that, heated i^2R means, dissipates heat by radiation and convection to a workpiece.

Metallic heating elements are alloys based on iron or on nickel for routine applications. Refractory alloys of metals such as platinum are used for temperatures above 1250°C. Non-metallic elements of silicon carbide and graphite are also used for high temperature (1600–3000°C) applications.

The metallic elements have a positive temperature coefficient of resistance, and hence for a particular applied voltage they settle to a stable operating temperature. Non-metallic elements have a negative temperature coefficient of resistance, and therefore the possibility of thermal runaway exists. Temperature control for all types of elements is achieved by varying the applied voltage, usually by thyristor switching.

9.3 Induction heating of metals

In induction heating, the metal to be heated (the workpiece) is inserted into a water-cooled coil that is supplied with an alternating voltage. The magnetic flux produced in the coil causes heating of the workpiece through induced currents and, if the workpiece is magnetic, through hysteresis effects. The density of the induced current is greatest at the surface of the workpiece, reducing as the distance in from the surface increases. This phenomenon is known as the skin effect.

The depth of penetration has been quoted (British National Committee *b*) as

$$d = 500\left(\frac{\varrho}{\mu f}\right)^{1/2} \tag{9.1}$$

where ϱ is the resistivity of the workpiece (Ωm), μ is the relative permeability of the material of the workpiece and f is the applied frequency (Hz).

Temperature control is achieved largely by choice of frequency and time of heating. High frequencies result in shallow penetration causing only the surface layers of the workpiece to be heated. Such a strategy is ideal for the surface hardening of certain steel tools. For heating the workpiece throughout, a lower frequency will need to be used than for localised surface heating. In applications involving plain steel workpieces the Curie point of the material of the workpiece is important, and if, as occurs often, the requirement is to heat the workpiece to the Curie point (about 750°C), it is easy to design a system that achieves this since, once the Curie point is exceeded, heating becomes much less efficient.

By suitable design of coils, induction heating can be made to produce very localised heating of parts of a workpiece. Further, such localised heating can easily be integrated within a production line so that, for instance, carbide tips may be brazed onto coal-cutting tools as they progress along a production line. Without the use of induction heating, the whole coal cutter would have inevitably been heated up while the brazing was performed.

9.4 Dielectric heating

In dielectric heating the material to be heated (the workpiece) is made the dielectric in a capacitor supplied by an alternating voltage. The heat generated in the workpiece depends on the applied voltage, the applied frequency and the loss angle of the material of the workpiece, considered as a dielectric, according to the equation

$$\text{power density} = kfV^2\varepsilon_r \tan\delta \tag{9.2}$$

where k = constant of proportionality
f = applied frequency
V = applied voltage
ε_r = relative permittivity of the material of the workpiece
δ = loss angle of the material of the workpiece

The loss angle varies very widely for different materials. Further, it is a function of both temperature and applied frequency, making precise calculation of the heating effect very difficult. However, water has a very high loss angle – much higher than almost any other material that is likely to be encountered. Thus dielectric heating will be very effective in heating any material that contains water. Further, dielectric heating is ideally suited to drying a workpiece that initially contains heterogeneously distributed moisture. As drying proceeds, the heating power will automatically be concentrated on regions where moisture still remains, while the bulk of the workpiece, being much less efficiently heated, is unlikely to be overheated.

For materials with high loss angles heated at high frequencies most of the heat energy is liberated in the surface layers of the workpiece, resulting in non-uniform heating. The penetration depth may be increased by lowering the applied frequency. However, the frequencies that may be used are fixed by international agreement in four bands centred at 13·56, 27·12, 915 and 2450 MHz. Further reading on principles and techniques of dielectric and microwave heating may be found in Section 9.7.

9.5 Control of steel melting in the electric arc furnace (Fig. 9.4)

An electric arc is a column of ionised gas that typically carries a current of 50 kA or more. The arc has a temperature of about 1400°C and dissipates its energy approximately equally by convection and radiation mechanisms.

9.5.1 Principle of electric arc when used for heating a molten metal bath

The arcs are established by driving the electrodes downwards until a short-circuit current occurs, limited by circuit impedance. The electrodes are then raised slowly until desired arc conditions are obtained.

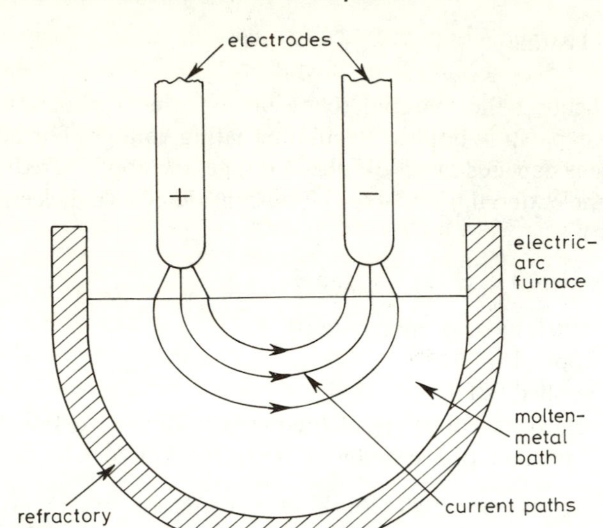

Fig. 9.4 *Electric arc furnace*

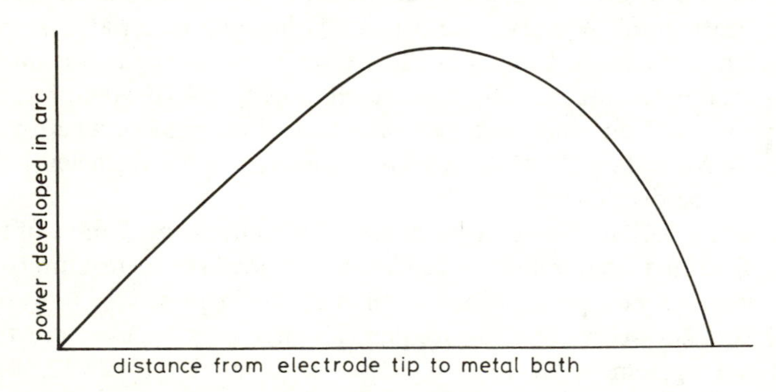

Fig. 9.5 *Power generated in arc for constant applied voltage as electrode–bath distance is varied*

The arc is a highly nonlinear phenomenon, and most large-scale melting furnaces use 3-phase alternating supplies with three electrodes arranged symmetrically. Under these conditions, the circuit reactance becomes important, and additionally, significant harmonic currents are produced. Thus the calculation of power generated by the arc is a non-trivial problem.

The power dissipated at the arc for a particular applied voltage is a nonlinear function of the distance from the electrode tip to the metal bath. Varying this distance varies both the voltage across the arc and the arc current. As the electrode tip to metal bath distance is varied (with constant applied voltage) the power generated in the arc varies as shown in Fig. 9.5.

From what has been said so far, it is already clear that control of the power input to an electric arc furnace will require some sort of accurate position servo to control the electrode–bath distance to a desired value. When it is considered that, in the early stages of melting down, the bath surface is moving rapidly as large pieces of solid-metal charge fall randomly, the task of the position-control system can be seen to embody a considerable dynamic element in addition to the obviously required steady-state accuracy. To give an approximate quantitative appreciation of the task, it may be of interest to note that an electrode with its associated hardware will weigh some 30 tonnes, that the velocity of lift may need to be 0·1 m/s with a required time to accelerate from rest to full speed of 0·2 s. During the high-performance control of the small electrode–bath gap the electrodes must not dip into the bath, for then the bath will absorb carbon from the graphite of the electrodes.

Of course, the electrode–bath distance cannot be measured directly, so the position-control loop operates to maintain actual power equal to desired arc power.

9.5.2 Measurement and control of temperature

Metallurgical considerations dictate that metal temperature should be closely controlled. In addition, the temperature of furnace linings must be prevented from reaching excessive temperatures to avoid shortening furnace life.

The main control of bath temperature is by careful programming of the electrical-energy input having regard to the (accurately known) weight of metal initially charged. Voltage is not usually continuously variable, but is instead

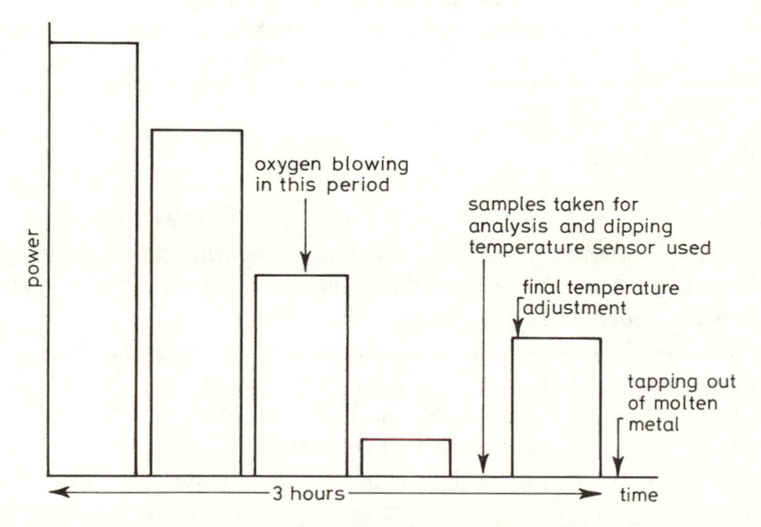

Fig. 9.6 *Typical power/time graph for an arc furnace melting 100% solid charge*

controlled between a few discrete levels by changing transformer tappings. However, the situation is nearly always complicated by the use of oxygen blowing. This is undertaken to speed the refining process, but the exothermic reactions involved increase the bath temperature by an acount that is difficult to calculate precisely. Fig. 9.6 shows a typical power/time graph for an arc furnace melting solid-metal scrap to reach the tapping-out point.

The temperature of the molten metal may be measured on a limited number of occasions during each batch by the use of a 'dipping thermocouple'. This is a long probe with an expendable end which is dipped through the surface layer of slag into the metal bath below to yield a measurement of bath temperature.

Avoiding excessive temperatures in the furnace linings, including the roof, is a considerable problem. First, it is difficult to measure accurately the temperatures of interest although there are a number of available devices for embedding in the refractory materials of which the linings are made. It is difficult to avoid localised overheating, due for instance to high radiation levels from one of the three arcs. This can arise because few, if any, 3-phase arrangements to supply the electrodes result in equal reactance in each phase. The control action to be taken when lining temperatures approach their permissible upper limit is, in the short term, to reduce the input power temporarily until lower lining temperatures are obtained. In the longer term, the furnace can perhaps be controlled to produce equal radiated power at each electrode to ensure more symmetrical heating of the lining.

Several arc furnaces working at a single location constitute a significant electrical load equal to that of a medium-sized town, and the production has to be carefully phased to avoid excessive instantaneous demands on the electrical supply. In addition, the nonlinear nature of the arc causes troublesome harmonics to be fed back into the public electricity supply.

Here, as in so many other applications, temperature control is embedded firmly in a set of other closely related control requirements.

9.6 Infra-red heating

Infra-red heating over the waveband 1–4 microns is used extensively in industry for providing localised heating and for drying and curing paints during production. Electrical infra-red sources include infra-red lamps, tubular heaters and narrow-beam concentrating radiators.

The wavelengths at which water vapour absorbs radiation strongly have also been considered in Chapter 5. It is clear that, for drying processes, it will be desirable to use an infra-red source whose waveband covers those wavelengths where water vapour is most absorbant of radiation.

In a typical industrial process line, products to be heated pass along a moving conveyor through an arch of infra-red radiators. Closed-loop control is quite difficult to achieve under these circumstances, since the temperature of the products is hard to measure accurately. On–off control of the radiators is not acceptable, since most types of radiator cool very rapidly. Even continuous control of radiator power is not very satisfactory, since any reduction in heater power alters the wavelength of the radiation.

9.7 References and further reading

British National Committee for Electroheat-Publications (BNC):
(*a*) 'The arc furnace'
(*b*) 'Induction heating applications'
(*c*) 'Induction heating equipment'
(*d*) 'Heat transfer for induction heating'
(*e*) 'Dielectric heating for industrial processes'
(*f*) 'The application of electric infra-red heating to industrial processes'
CUBE, H. L. V. (1981): 'Heat pump technology' (Butterworth)
HEAP, R. D. (1983): 'Heat pumps' (E and FN Spon)
REAY, D. A., and MACMICHAEL, D. B. A. (1979): 'Heat pumps: design and applications', (Pergamon, Oxford)

Chapter 10

Temperature control – 1: On-off control

10.1 Characteristic features of temperature-control problems

(i) In almost all temperature-control applications, complex heat-transfer mechanisms are involved that are difficult to model accurately.
(ii) Heat supply and heat removal occur by different means, resulting in an asymmetry of behaviour of a controlled system.

For lumped processes (i.e. for those processes where the basic aim is to control the mean temperature within some spatial region):

(iii) Processes may be represented by predominantly first-order dynamics. This means that high loop gains may be implemented without danger of instability. It also follows that many control loops may be implemented satisfactorily by high-gain strategies, notwithstanding lack of quantitative knowledge of the process.
(iv) Temperature measurements tend to be available only at one or, at most, a few discrete points in space, so that the temperature to be controlled often differs from the temperature that can be measured.

For distributed parameter processes (i.e. for those processes where the basic aim is to control the temperature distribution within some spatial region):

(v) It will usually be necessary to obtain a good quantitative model of the process before a controller can be designed. Frequently flow and/or diffusion mechanisms will need to be incorporated into the model, which, as a result, may become large and unwieldy.
(vi) Only rarely will scanning or distributed temperature sensors give an accurate measurement of the temperature distribution that is of interest. In most cases, the process model described in (v) will have to be called upon to act as an interpolating function between the measured data that are available.

10.2 Temperature control in terms of energy balance

The rate of heat loss from a heated body is always an increasing nonlinear function of temperature such as is shown in Fig. 10.1*a*.

Consider the problem of controlling the temperature of a volume of liquid

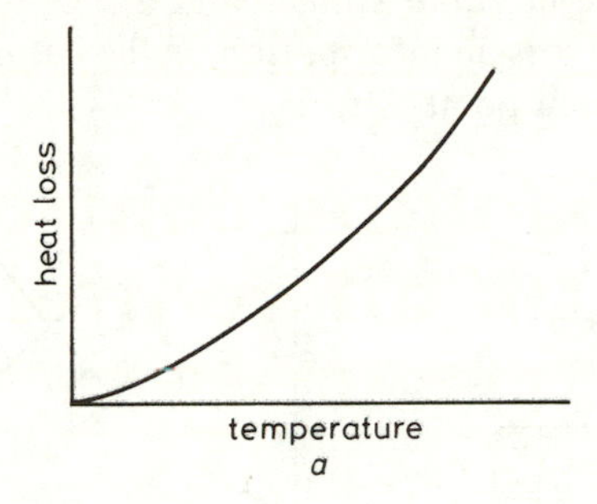

Fig. 10.1a *Heat loss as a function of temperature*

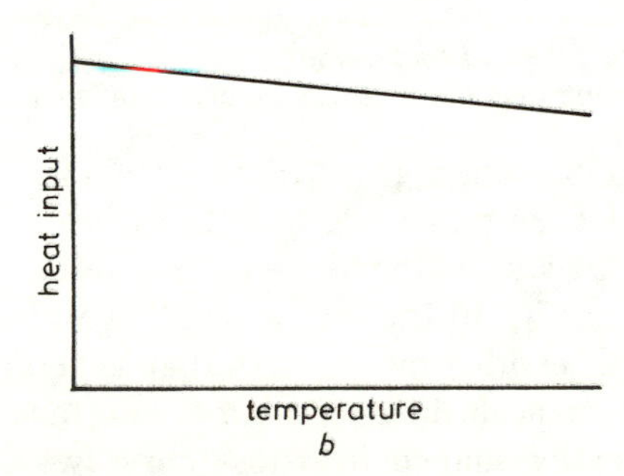

Fig. 10.1b *Heat input as a function of temperature*

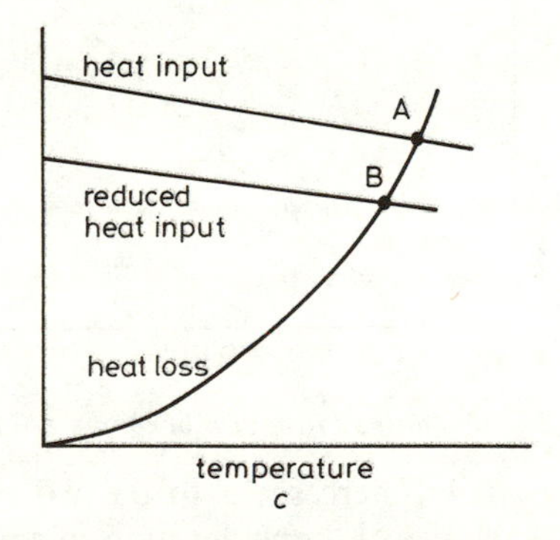

Fig. 10.1c *Equilibrium is reached at point A, or, with reduced heat input, at point B*

using a totally immersed electrical resistance heater. All the heat generated electrically must pass into the liquid, so that if a constant voltage is applied to the heater then there will be a nearly constant heat input into the liquid over a wide range of temperature (see Fig. 10.1*b* where the slight fall in heat input with rising temperature allows for increasing resistance of the heater element with rising temperature).

When the curves are superimposed as in Fig. 10.1*c*, they indicate that the process will be expected to settle at the working point A. If the input voltage is reduced, a new heat-input curve applies and a new working point B will be reached. Notice that there is no information on the nature of the time behaviour as the system moves from point A to B.

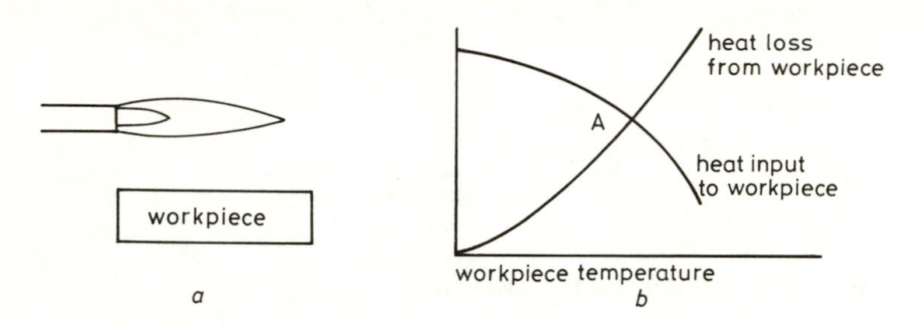

Fig. 10.2 *(a) Workpiece is heated by a flame*
(b) Heat loss and heat input curves predict equilibrium at point A

Next consider the process in Fig. 10.2*a*, where a horizontal flame is radiating to a workpiece. For a fixed flame, the useful heat transfer rate will fall as the workpiece temperature rises and the heat-gain and -loss curves applicable to the workpiece will be as in Fig. 10.2*b*, with a steady working point at A.

Finally, consider an exothermic process that generates internal heat as a function of temperature according to a curve like that of Fig. 10.3. In conjunction with a normally shaped heat-loss curve we obtain the interesting

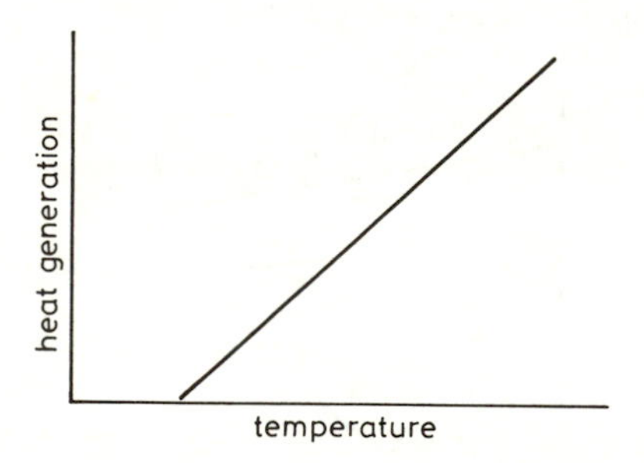

Fig. 10.3 *Exothermic heat generation as function of temperature*

situation shown in Fig. 10.4. There seem to be two possible steady working points, marked A and B. However, consideration of small perturbations about point A shows that it is unstable. For temperatures below θ_A the process will cool right down to ambient, whereas once temperature θ_A is exceeded the process will run off to the stable operating point B. Assume, as is often the case, that the process must be operated at some given temperature between θ_A and θ_B, then control consists in first supplying heat until the temperature θ_A is exceeded and then applying controlled forced cooling so that the desired steady temperature is maintained.

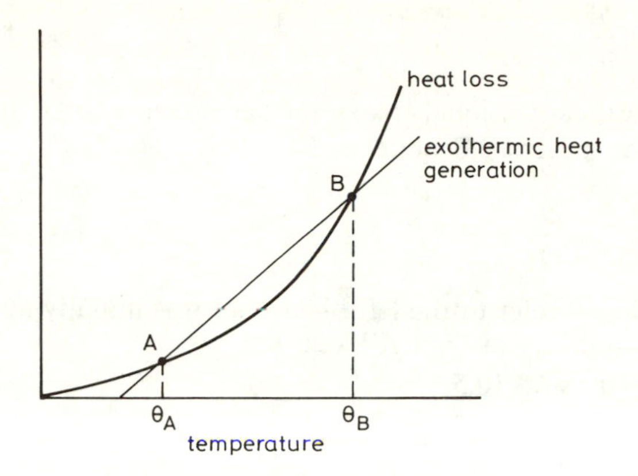

Fig. 10.4 *Of the two apparently possible equilibrium prints, only point B is stable*

10.3 Elementary models of thermal processes

In the following development, units are deliberately omitted to avoid obscuring the simple structure.

Fig. 10.5 shows a block of material of mass M that is at a uniform temperature θ_0. Suppose that the temperature of the block is somehow raised to the

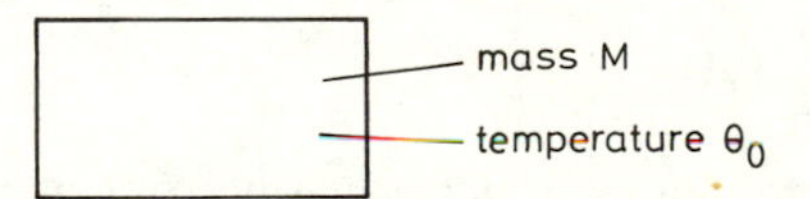

Fig. 10.5 *An idealised block of material at a uniform temperature*

uniform temperature θ_1, where $\theta_1 > \theta_0$, then the increase Δq in internal energy in the block is given by

$$\Delta q = C(\theta_1 - \theta_0) = MC_p(\theta_1 - \theta_0) \quad (10.1)$$

where C is the thermal capacitance of the block, given by MC_p, where C_p is the specific heat of the material of the block over the range θ_0 to θ_1. Alternatively, we can write eqn. 10.1 in the form

$$dq = MC_p d\theta. \quad (10.2)$$

Next consider the arrangement of Fig. 10.6 where two blocks of metal at temperatures θ_1, θ_1, respectively, are separated by an insulating barrier of thickness b. The rate of heat flow through the insulating barrier is given by

$$\frac{dq}{dt} = \frac{1}{R}(\theta_1 - \theta_0) \tag{10.3}$$

where R is the thermal resistance of the insulating barrier, given by

$$R = \frac{b}{ae} \tag{10.4}$$

in which a is the cross-sectional area of the barrier and e is the thermal resistivity of the insulating material. From eqn. 10.2,

$$\frac{dq}{dt} = C_0 \frac{d\theta_0}{dt} \tag{10.5}$$

where the suffices 0 refer to the left block that was initially at temperature θ_0. (C_0 is the heat capacity of the left block.)

From eqns. 10.3 and 10.5,

$$\frac{1}{R}(\theta_1 - \theta_0) = C_0 \frac{d\theta_0}{dt} \tag{10.6}$$

or

$$\frac{d\theta_0}{dt} + \frac{1}{RC_0}\theta_0 = \frac{1}{RC_0}\theta_1 \tag{10.7}$$

with solution

$$\theta_0(t) = \theta_0(0) + (\theta_1 - \theta_0(0))(1 - e^{-t/RC_0}) \tag{10.8}$$

or taking Laplace transforms of eqn. 10.6,

$$SC_0\theta_0(s) + \frac{1}{R}\theta_0(s) = \frac{1}{R}\theta_1(s) \tag{10.9}$$

where we have neglected initial conditions since our objective is to proceed to a transfer function-representation.

From eqn. 10.9

$$\frac{\theta_0(s)}{\theta_1(s)} = \frac{1/R}{SC_0 + (1/R)} = \frac{1}{1 + SC_0R} \tag{10.10}$$

Eqn. 10.10 shows clearly the essential first-order dynamics of the heat transfer from the heat source to the isolated block. The product C_0R may be considered to be the time constant of the process.

Practical, as opposed to idealised, thermal processes involve complex physical boundaries, fluids in motion, non-uniform temperature distributions and non-linear heat-loss terms. Thus precise models of real thermal processes are in general very complex. Such models may be useful to process designers, but for control-system design, the requirement is for much simpler dynamic models for which eqn. 10.10 often forms the starting point.

10.4 On–off or high–low control of temperature

Many temperature-control loops operate quite satisfactorily under on–off or high–low control.

For most applications, a 2-position relay acts as the controller, applying full heating power when the measured temperature is low, and either no heating power or low heating power when the measured temperature is high (Fig. 10.7). The advantages of this form of control are:

(i) The controller is no more than an error-driven switch. As such it is cheaper than a controller that has a continuously variable output.
(ii) More significantly than (i), on–off actuators, such as solenoid valves are nearly always considerably cheaper than their continuously variable counterparts.
(iii) Temperature control through on–off relays is adequate for many applications.

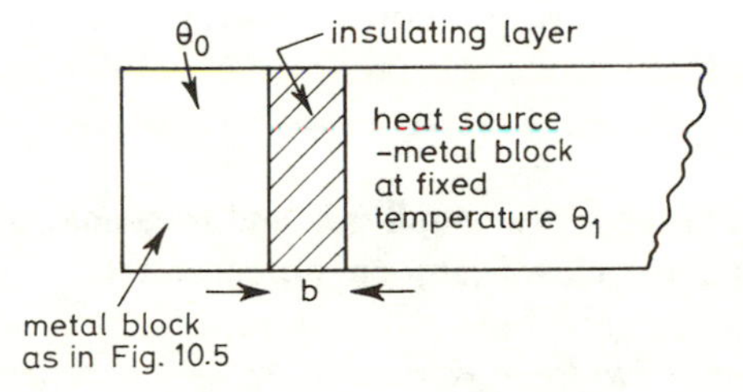

Fig. 10.6 *The right block is considered to be an ideal heat source; i.e. it remains at temperature θ_1, despite loss of energy*

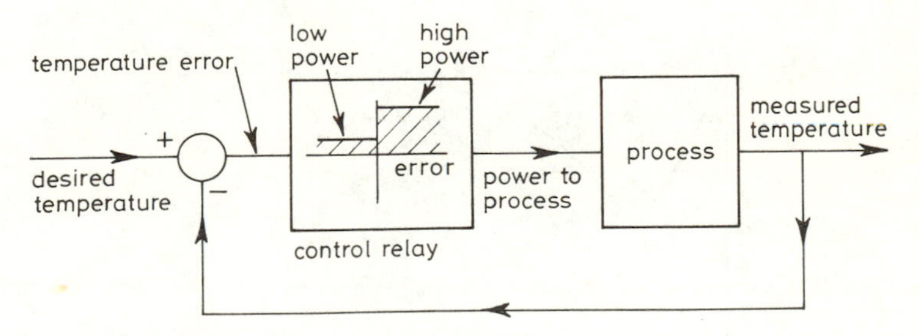

Fig. 10.7 *Simple high–low control of temperature*

10.5 High–low control of fuel-fired furnaces

On–off control of a fuel-fired furnace, if carried out literally, would imply periodic extinguishing and re-ignition of flames. (Where the fuel is coal, different difficulties arise from on–off control and these were discused in Section 8.14.1.) Instead of on–off control, high–low control is used using the configuration in

Fig. 10.7. An on–off actuator is used in the main fuel line while a permanently open bypass provides continuity of ignition.

The attractions of the high–low system are that it will give good or adequate temperature control in many applications and the on–off actuator is much cheaper than a continuous actuator of the same rating.

Disadvantages of the high–low system are associated not so much with the temperature control itself but rather with pressure control and furnace-atmosphere control. Clearly, the pressure-control system has to contend with step changes in conditions, but, most important, experience of many industrial applications has shown that it is very difficult to maintain a desired composition of gases inside the furnace (atmosphere control) under very low fuel-input conditions. Because of the two effects just described, the control engineer will often increase the flow rates of the 'low' setting as far as he dares, so that the furnace remains in a reasonable combustion rate at all times. This setting must be done with care so that at no time can the furnace temperature rise out of control. Of course, as a fail-safe additional feature, all such high–low schemes must be backed up by an emergency control switch that shuts the furnace down if over-temperature does occur.

10.6 Approximate analysis of the amplitude and frequency of oscillation of a heating process under closed-loop on–off control

Development of a simple model to represent the on–off behaviour

Assume initially that the process is first order with time constant T_1 for heating and time constant T_2 for cooling (Fig. 10.8).

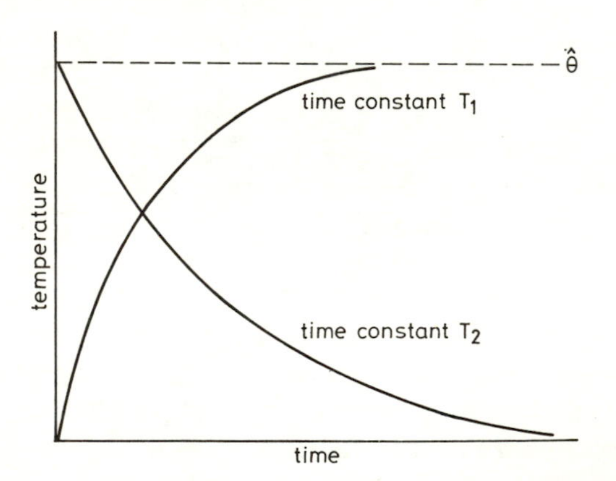

Fig. 10.8 *Heating and cooling curves*

Without loss of generality we take ambient temperature as 0°C and the maximum temperature reached by heating as $\hat{\theta}$°C. Then, during continuous heating from time zero, the temperature at any time t is given by

$$\theta(t) = \hat{\theta}(1 - e^{-t/T_1}) \tag{10.11}$$

and

$$\dot{\theta}(t) = \frac{1}{T_1}\hat{\theta}e^{-t/T_1} \tag{10.12}$$

to express $\dot{\theta}(t)$ in terms of $\theta(t)$ rather than of t we substitute eqn. 10.11 into eqn. 10.12 and obtain:

$$\dot{\theta}(t) = \frac{1}{T_1}(\hat{\theta} - \theta(t)) \tag{10.13}$$

Similarly, during cooling,

$$\theta(t) = \hat{\theta}e^{-t/T_2} \tag{10.14}$$

$$\dot{\theta}(t) = \frac{-1}{T_2}\hat{\theta}e^{-t/T_2} \tag{10.15}$$

and combining these two equations yields

$$\dot{\theta}(t) = \frac{-1}{T_2}\theta(t) \tag{10.16}$$

For small deviations about a nominal (desired) temperature θ_d, we can consider the heating and cooling to follow constant ramps of slope

$$\frac{1}{T_2}(\hat{\theta} - \theta_d), \quad \frac{-1}{T_2}\theta_d, \quad \text{respectively}$$

The response of such a system to alternate (unequal) periods of heating and cooling is sketched in Fig. 10.9 for two different mean temperature levels, assuming that the heating and cooling time constants, T_1, T_2 are equal.

The simple analysis we have just completed gives some understanding of the response to be expected under on–off control. However, there is a disappointment. If we try to determine the amplitude and frequency of the temperature signal under ideal on–off control, we obtain zero amplitude and infinite frequency. To model the oscillatory behaviour that we know from physical intuition must exist during on–off control, it is necessary to include a delay term and/or higher-order terms in the model of the heating process. We can argue that every physical control switch will have some hysteresis whose effect may be represented by assuming that switching occurs not when $\theta = \theta_d$ but rather when $\theta = \theta_d + E$ (temperature rising) and when $\theta = \theta_d - E$ (temperature falling).

The temperature-measuring sensor will always have some dynamic effect associated with it, although usually the speed of response will be at least an order of magnitude faster than that of the process. The effect of such rapid measurement dynamics will be to replace the sharp corners on the waveform of Fig. 10.9 by smooth corners, as well as to move the triangular waveform to the right just as would some equivalent time delay.

Finally, recall that the first-order lumped-parameter approximations of heat transfer are simplistic representations of complex heat-transfer phenomena that, ideally, should be modelled by sets of partial differential equations. In particular, the lumped approximation fails to allow properly for the time taken for heat to diffuse. The first-order approximation may often be improved by incorporating a suitably chosen delay term to allow for the diffusion time.

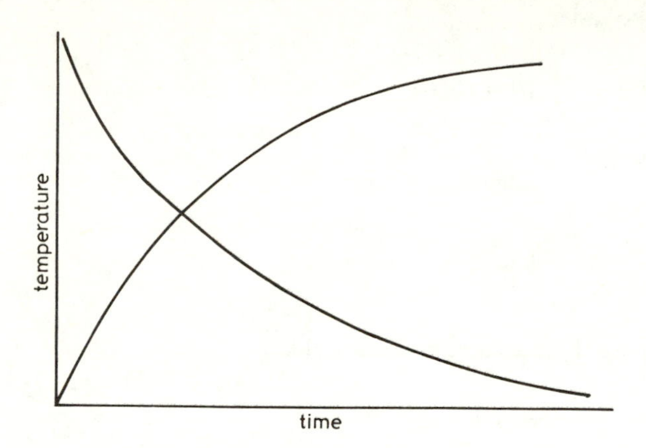

Fig. 10.9 *Response of a system to alternate periods of heating and cooling*
Two different mean temperatures illustrated

To summarise, the basic first-order model needs to be augmented to agree with the real heating processes:

(i) By incorporating fast acting dynamics (nominally representing the measurement sensor) to 'round the corners' of the response shown in Fig. 10.9.
(ii) By incorporating a delay term to take account of heat diffusion times and to allow for any significant hysteresis in the control relay.

We arrive at the simple model

$$G(s) = \frac{K_1 e^{-ST_d}}{(1 + ST_1)(1 + ST_2)} \tag{10.17}$$

in which

$$K_1 = K_{11} \text{ (heating)} \qquad T_1 = T_{11} \text{ (heating)}$$
$$= K_{12} \text{ (cooling)} \qquad = T_{12} \text{ (cooling)}$$

T_1 = predominant time constant of the heating/cooling process
T_2 = short time constant associated with sensor dynamics
T_d = delay time that represents distributed-parameter effects, neglected higher-order effects and any hysteresis in the control relay.

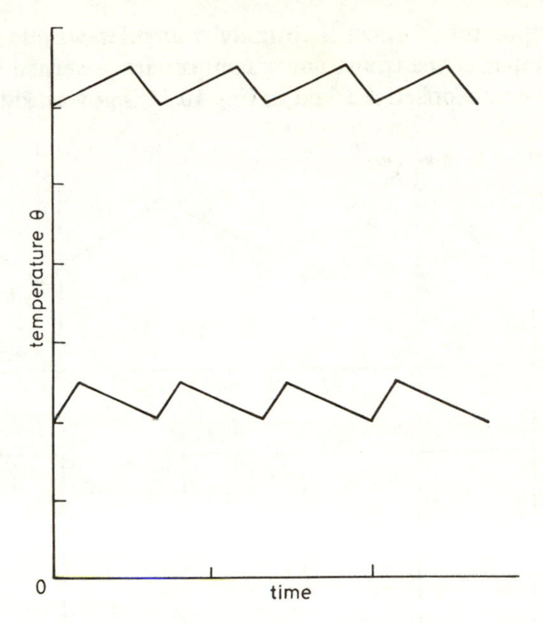

Fig. 10.10 *Process heating and cooling curves*

10.7 Obtaining numerical coefficients for the approximate model from simple tests

We postulated above that the process may be modelled by an expression of the form

$$G(s) = \frac{Ke^{-sT_d}}{(1 + sT_1)(1 + sT_2)} \tag{10.18}$$

where T_1 is the main time constant of the process (it may take different values for heating and for cooling), T_d is a delay time that represents the dynamic effects of heat diffusion that are otherwise neglected in a lumped-parameter approximation. T_2 represents, notionally at least, the relatively short time constant of the temperature-measuring sensor. (More pragmatically, it is included, because, without it, the model response cannot be made to match the observed response of an actual heating process.)

10.7.1 Initial test

First we obtain, if possible, temperature/time records of the process during steady heating and during steady cooling (Fig. 10.10).

From the curves, we select a temperature where heating and cooling rates are approximately equal. We now apply a rectangular waveform with equal time for power-on and power-off until the process is cycling repeatedly as shown in Fig.

10.11. The temperature follows a roughly triangular waveform with rounded vertices. The slopes of the triangular waveform are governed by the main time constant T_1. The time offset, marked in Fig. 10.11 is governed by a combination

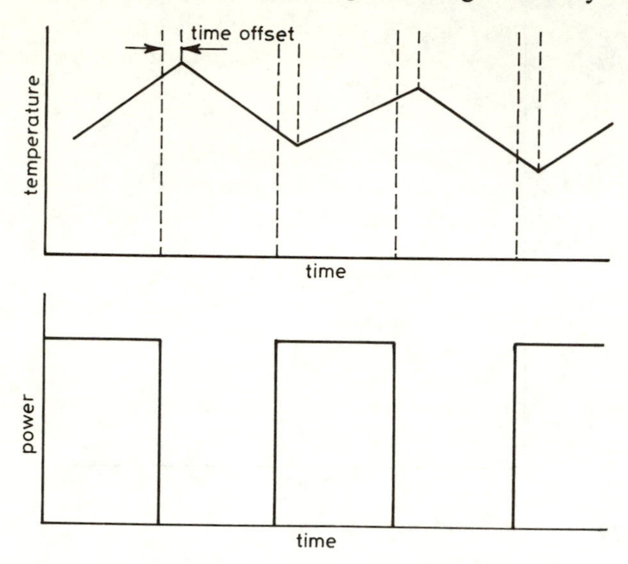

Fig. 10.11 *Process response to a rectangular input waveform*

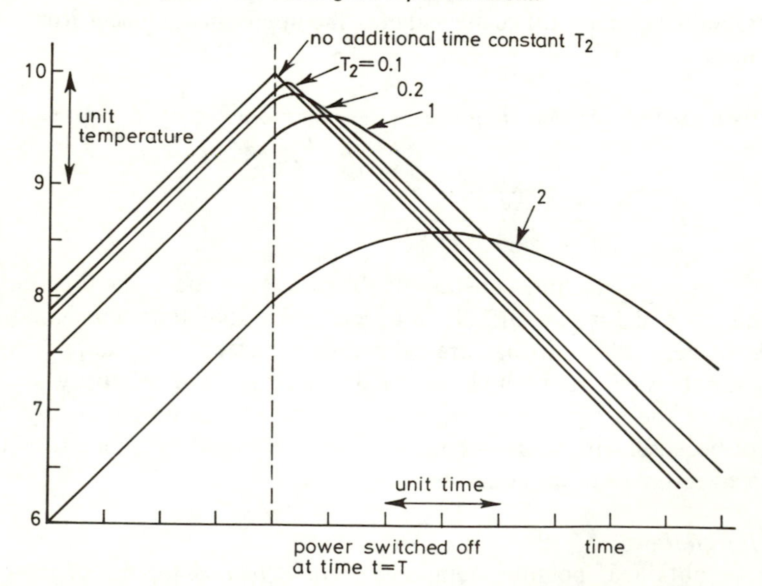

Fig. 10.12 *Sample solutions of eqs. 10.18 with normalised scales to give unity slopes*
With T_d = 0, for various secondary time constants T_2

of T_d and T_2. The roundedness of the vertices is governed by the time constant T_2. To obtain a quantitative understanding of the interrelation of T_d and T_2, Fig. 10.12 should be examined. It shows the response on axes scaled so that the triangular waveform has unity slope, for a number of different values of the time constant T_2. It can be seen that, as T_2 increases, so the time offset increases but the peak amplitude decreases. If the records of tests carried out are plotted on scales chosen to give unity slope to the waveform, then an estimate may be made of the time constant T_2 and the delay time T_d for insertion in a model of the form given in eqn. 10.18.

10.7.2 How the curves of Fig. 10.12 were obtained

Assume that the basic triangular waveform generated by the on–off rectangular application of power has been scaled to have unity slope. The waveform is now modified by the dynamic element

$$\frac{1}{1 + sT_2}$$

Let the applied triangular waveform have period T, then it can be represented (locally) by the expression

$$u(t) = t - 2t\,h(t - T)$$

where $h(t - T)$ is a step function, starting at $t = T$,

$$u(s) = \frac{1}{s^2}(1 - 2e^{-sT})$$

and the Laplace transform of the waveform after interaction with the $1/(1 + sT_2)$ element is

$$y(s) = \frac{1}{s^2}\frac{1}{1 + sT_2} - \frac{2e^{-sT}}{s^2(1 + sT_2)}$$

for $t < T$, we obtain

$$y(t) = T_2\left(e^{-t/T_2} + \frac{t}{T_2} - 1\right)$$

and for T_2 small, the ascending ramp tends to the expression

$$y(t) = (t - T_2)$$

for $t \geqslant T$, we obtain

$$y(t) = (t - T_2) - 2T_2\left(e^{-(t-T)/T_2} + \frac{(t - T)}{T_2} - 1\right) \tag{10.19}$$

and this equation defines the shape of the descending ramp.

10.8 Prediction of the amplitude and frequency of oscillation under on–off control, using the simple model of the heating process

Assume that the process, modelled by eqn. 10.18, is under on–off control in closed loop. The response will have the form shown in Fig. 10.13. The waveform is basically triangular with vertices rounded by the first-order dynamics in the model. We are interested to predict the amplitude and frequency of the waveform and also the position of the desired temperature line in relation to the waveform, so that we can estimate the temperature error as a function of time.

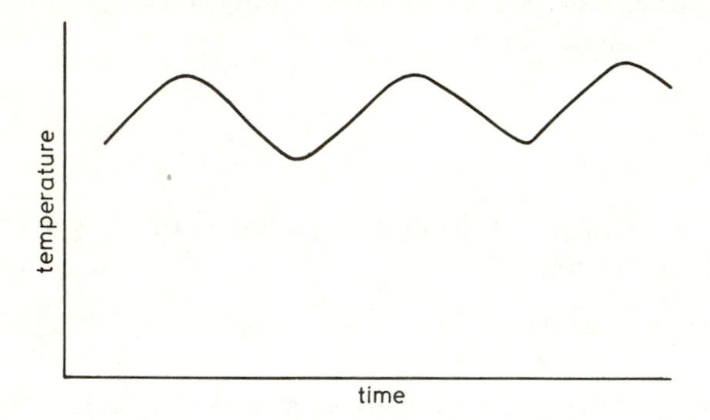

Fig. 10.13 *Typical temperature/time response under on–off control*

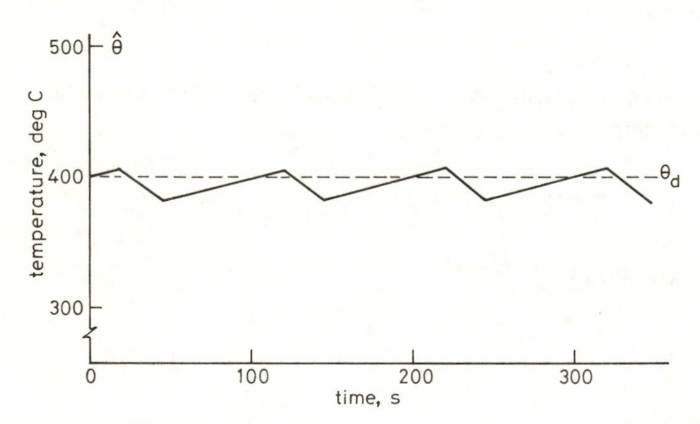

Fig. 10.14 *Expected behavior of the model process under closed-loop on–off control*

Let $\hat{\theta} = 500°C$, $\theta_d = 400°C$, $T_d = 20\,s$, $T_1 = T_2 = 400\,s$ then Fig. 10.14 illustrates the expected behaviour under closed-loop on–off control. The period of the oscillation is 100 s, maximum overshoot = 5°C, maximum undershoot = 15°C. (Fig. 10.15 shows the information from Fig. 10.14 on an amplified scale.)

In general, the period of the oscillation is given by

$$2T_d + \left(\frac{\theta_d T_d}{T_2}\bigg/\frac{\hat{\theta} - \theta_d}{T_1}\right) + \left(\frac{(\hat{\theta} - \theta_d)T_d}{T_1}\bigg/\frac{\theta_d}{T_2}\right) \tag{10.20}$$

and in the special case where $\theta_d = \hat{\theta}/2$ and $T_1 = T_2$, the period is

$$2T_d + \frac{\theta_d T_d}{T_1}\bigg/\frac{\theta_d}{T_1} + \frac{\theta_d T_d}{T_1}\bigg/\frac{\theta_d}{T_1} = 4T_d \qquad (10.21)$$

The period $4T_d$ is the minimum possible. It occurs when $T_1 = T_2$ and, in that case,

$$\text{maximum overshoot} = \text{maximum undershoot} = \theta_d \frac{T_d}{T_1}$$

10.8.1 Discussion

Where on–off control is to be implemented for a heating process for which heating and cooling time constants are approximately equal, best results will be achieved if $\hat{\theta}$ (the temperature that would be reached eventually with power fully on) is chosen according to the relation

$$\hat{\theta} = 2\theta_d \qquad (10.22)$$

In many cases, $\hat{\theta}$ is manipulable by a flow-control valve (oil or gas system), by choice of gear ratio (coal-fired furnace) or by choice of transformer tap (electric furnace), and therefore, eqn. 10.22 can be made to hold approximately. However, in many industrial applications, there is a need for the maximum possible rate of energy input at the start of a batch cycle, after a sudden increase in load or after a step increase in desired temperature. This implies that $\hat{\theta}$ should be as high as possible.

In general, $T_1 \neq T_2$ and the control system operates not between on–off but rather between high–low limits. Thus, there is another manipulable temperature $\check{\theta}$, being that temperature reached by the steady application of 'low' energy input. In making the choice of $\check{\theta}$, not only control aspects have to be considered but also furnace pressure and atmosphere aspects (roughly, if $\check{\theta}$ is chosen to be very low, it may prove difficult to maintain proper pressure and combustion conditions).

10.9 Prediction of process behaviour under on–off feedback control

Consider a process like that shown in Fig. 10.16. An on–off -controlled electrical resistance heater is used to heat a metal block to a desired temperature θ_d. The metal block is subject to cooling by liquid passing through embedded tubes. The liquid flow rate Q can vary over a wide range and is not under our control. The model for this process will, approximately, have the form

$$G(s) = \frac{Ke^{-sT_d}}{(1 + sT_1)(1 + sT_2)} \qquad (10.23)$$

in which K and T_1 will be strongly dependent on water flow. Neglecting the effect of the time constant T_2, we consider the behaviour of the process under on–off feedback control with different rates of liquid flow. Figs. 10.17, 10.18 and 10.19 show the expected behaviours under zero, medium and high liquid flows, respectively. Notice particularly how the extra heating load imposed by the high liquid flow is expected to cause the mean controlled temperature to fall below θ_d; i.e. note that even under on–off control there will be a temperature offset due to loading.

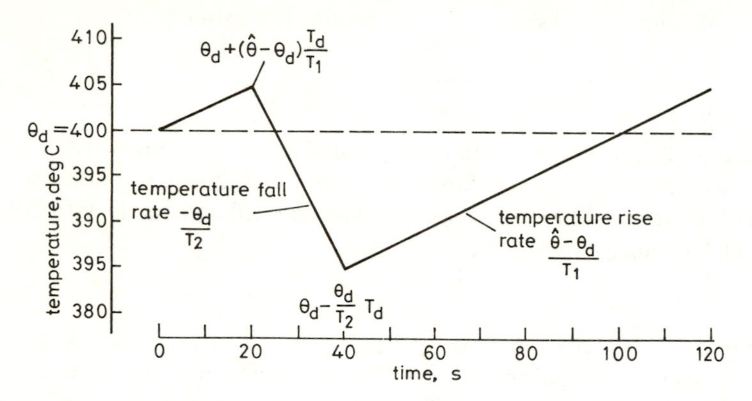

Fig. 10.15 *Expected behaviour of the process model under closed-loop on–off control* Amplified scales

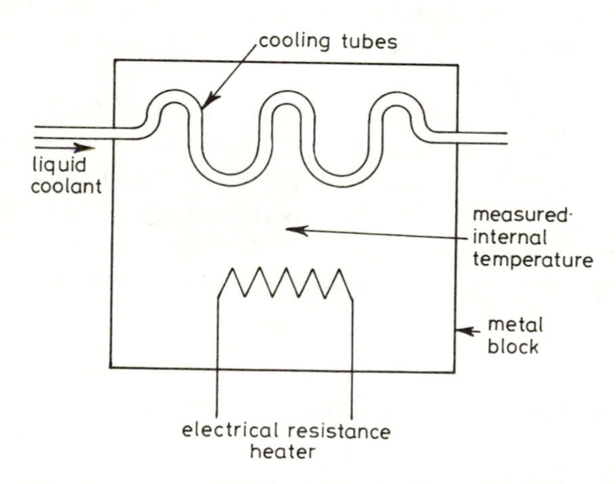

Fig. 10.16 *The metal block whose temperature is to be controlled*

To allow comparison of theoretical predictions with results actually obtained, an experimental device similar to that shown in Fig. 10.16 was set up and put under closed-loop on–off control for conditions of zero, medium and high liquid flow, producing results shown in Figures 10.20, 10.21 and 10.22 respectively. It can be seen that the experimental results agree with the results predicted by the simple model, particularly in regard to the temperature offset that occurs on application of the heating load. It should be noted that the quality of on–off control is seriously degraded when the heating and cooling rates are markedly different.

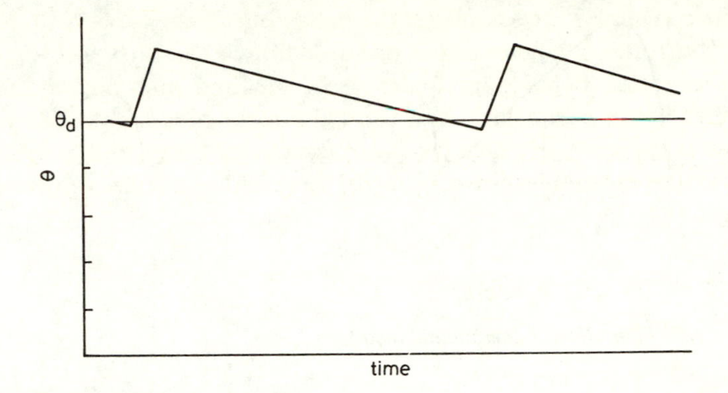

Fig. 10.17 *Expected control behaviour: zero liquid flow*

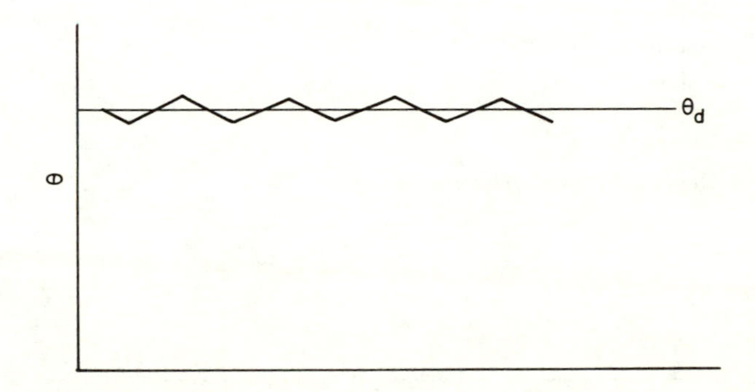

Fig. 10.18 *Expected control behaviour: medium flow rate*

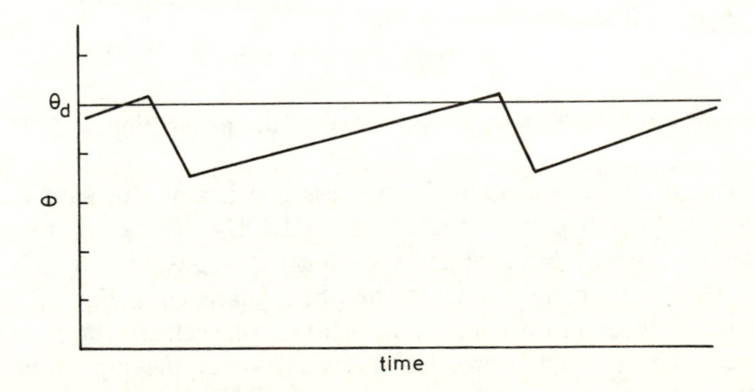

Fig. 10.19 *Expected control behaviour: high liquid flow*

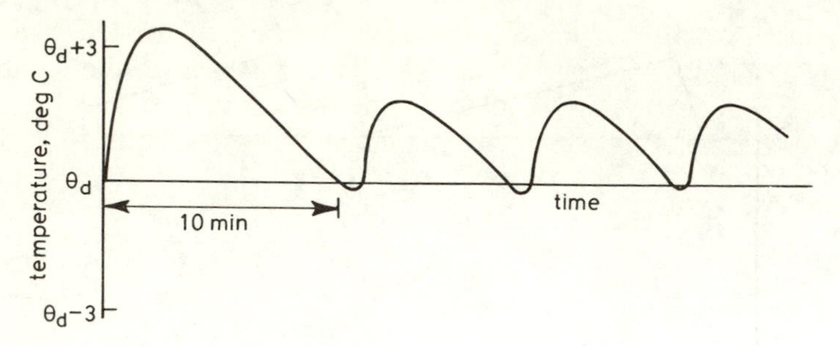

Fig. 10.20 *Zero liquid flow: experimental result*

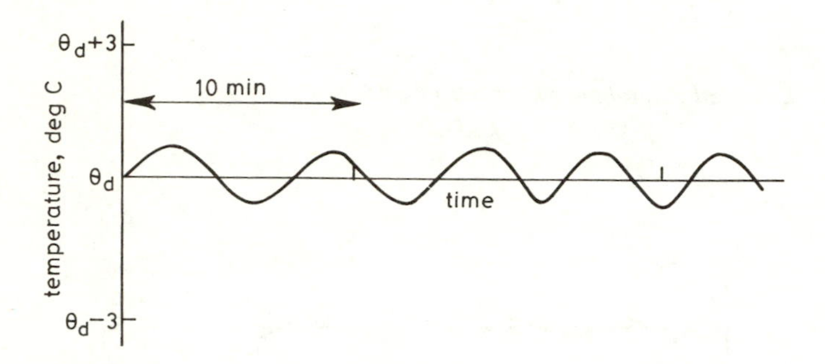

Fig. 10.21 *Medium liquid flow: experimental result*

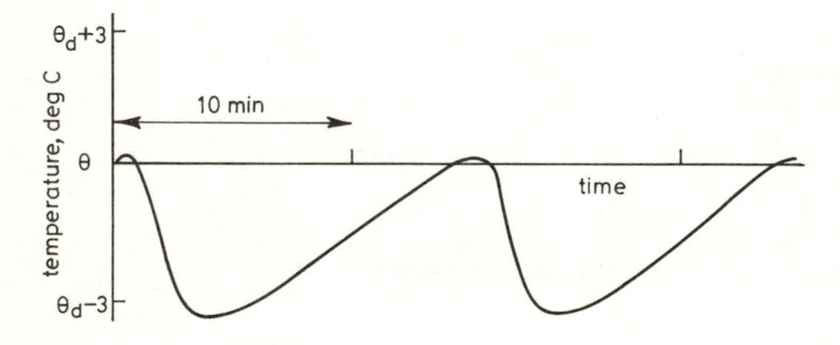

Fig. 10.22 *High liquid flow: experimental result*

10.10 Dynamics of on–off control, illustrated in the phase plane

In on–off control, the response of the process consists of alternate sections of heating curve and cooling curve (Figs. 10.23 and 10.24). Notice that the heating and cooling curves may have widely differing time constants. The behaviour may also with advantage be shown in the phase plane as in Fig. 10.25.

The phase plane diagram (Fig. 10.25) shows very clearly how the system cycles between power-on and power-off states. However, the important question soon arises: what determines the magnitude of $\Delta\theta$. (If this can be found, then other important variables such as the cycle time can be estimated.) The difficulty lies in the fact that using a first-order heating model and assuming an ideal on–off relay with no hysteresis, then, theoretically, $\Delta\theta = 0$ and the system should cycle at infinite frequency over a negligibly small temperature range.

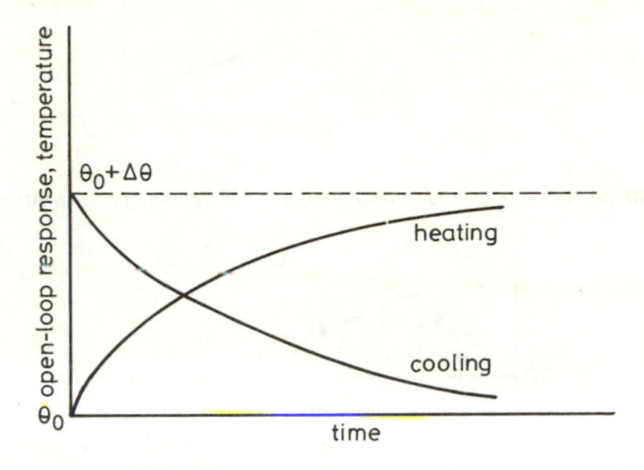

Fig. 10.23 *Heating and cooling curves for a typical process*

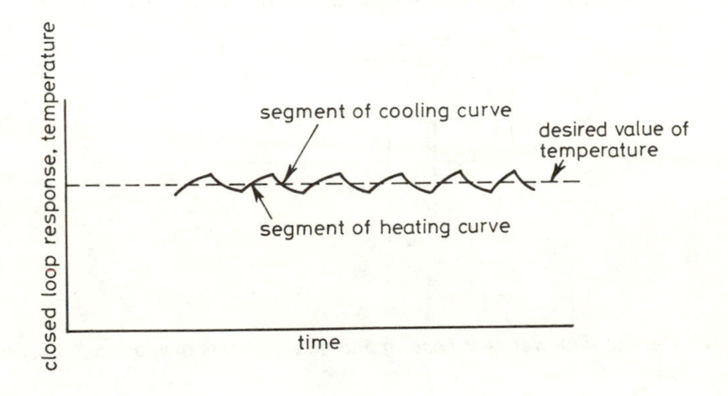

Fig. 10.24 *Typical closed loop on–off control for a typical process*

Assume that we could obtain an on–off relay with zero hysteresis (within the measurement accuracies that are important in most applications, such relays are available); thus we know from physical intuition that, if the temperature is rising with power on and we switch off exactly when $\theta = \theta_d$, then the temperature will go on rising for a time after the switch off. But this means that the heating process cannot be of first order, for if it were the temperature would start to fall instantaneously immediately on switch off. The important implication is that the cycling shown in Fig. 10.25 cannot be modelled with the idealised elements, hysteresis-free relay and first-order process equation. (Figs. 10.26–10.28).

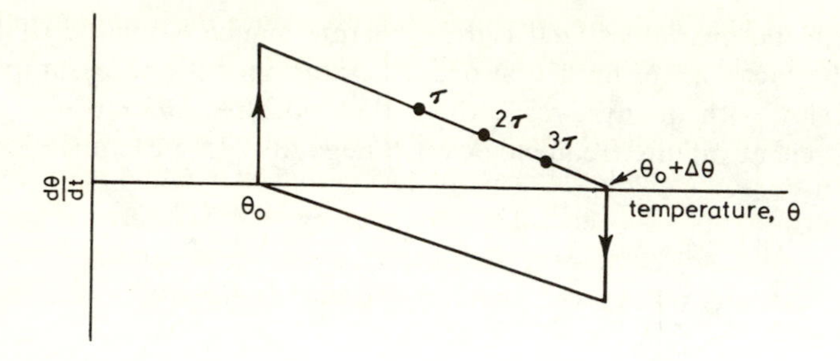

Fig. 10.25 *Phase-plane diagram of idealised on–off closed-loop temperature control*

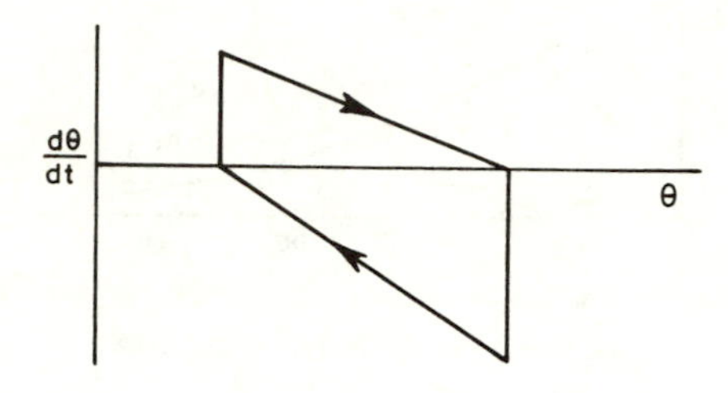

Fig. 10.26 *Cycling with different heating and cooling rates*

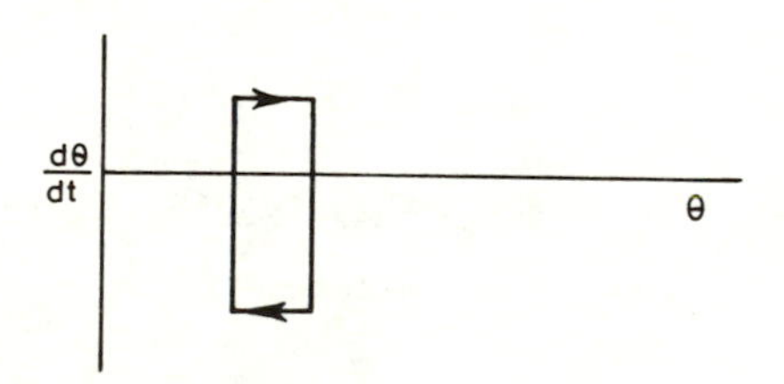

Fig. 10.27 *Cycling with different heating and cooling rates over a small temperature excursion*
(Such that rates of change may be considered constant)

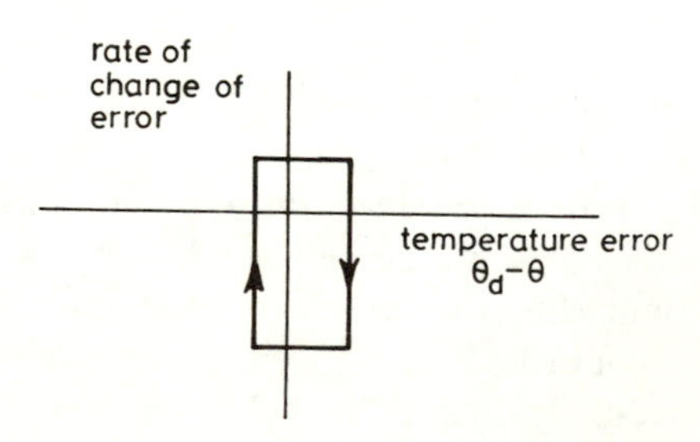

Fig. 10.28 *Typical cycling of error in an on–off temperature control system*

10.11 Use of phase-advance compensation to reduce the amplitude of oscillation under closed-loop on–off control

A phase-advance element, when incorporated into a closed-loop on–off system as shown in Fig. 10.29, advances the switching so that the relay changes over before zero error is reached. The two phase-plane diagrams (Figs. 10.30 and 10.31), in each case depicting the steady state, show clearly how the phase-advance element moves the switching point and reduces the amplitude of oscillation.

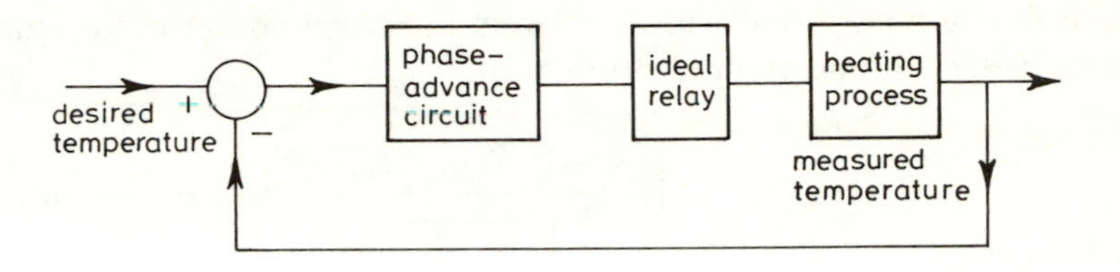

Fig. 10.29 *Incorporation of a phase-advance circuit into an on–off control loop*

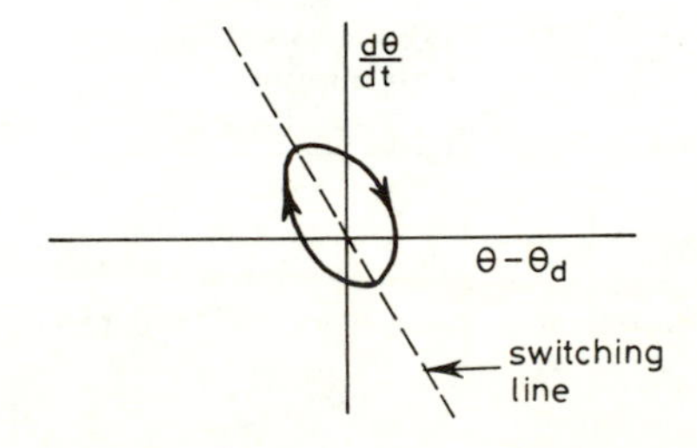

Fig. 10.30 *Phase-plane diagram of system performance with phase-advance circuit incorporated*

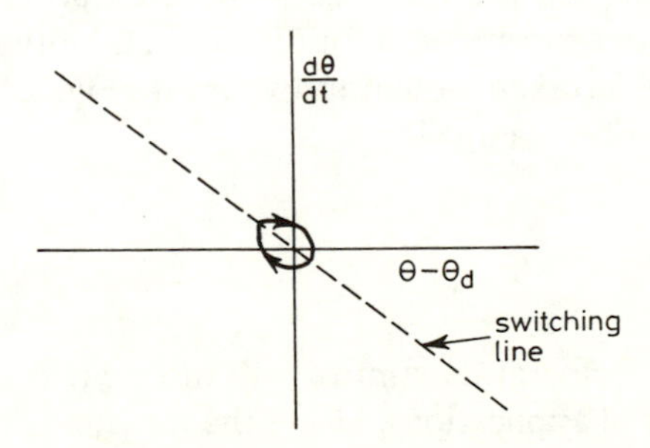

Fig. 10.31 *As Fig. 10.30 but with increased phase-advance action*

10.12 Use of a minor feedback loop to compensate for hysteresis in a control relay

All on–off control valves or relays exhibit hysteresis to some degree. If the effect of such control-valve hysteresis is expected to seriously degrade system performance, it may be worthwhile compensating for the effort by the use of a minor feedback loop.

Let the control-valve characteristic be as shown in Fig. 10.32: the logic block produces the signal $z = p$, when $u > 0$; $z = -q$, when $u < 0$. Obviously, perfect compensation can be obtained only if the values of p and q remain constant over a long period. However, even imperfect compensation may still be much better than no compensation at all.

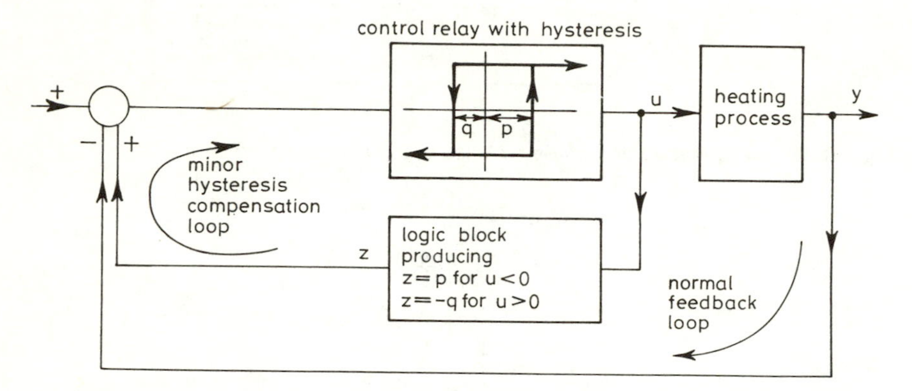

Fig. 10.32 *Illustrating the use of a minor feedback loop to compensate for control-valve hysteresis*

The dynamic behaviour of the minor loop must, however, be analysed with care to ensure that new problems are not created by the addition of compensation. One obvious unwanted, and even disastrous, effect would be that in which the minor loop was unstable, causing continuous chattering of the main valve that is supposed to be compensated. To avoid such problems, the minor loop must be designed with proper regard for dynamic stability requirements. Before the design can be undertaken a quantitative knowledge of the dynamics of the actuator will need to be obtained.

10.13 Conclusions

On–off closed-loop temperature control will often yield results that are quite adequate for industrial applications. Using the methods explained, an approximate process model may be derived from simple tests. The model may then be used to predict approximate values for the period and amplitude of the oscillation of temperature under closed-loop control, together with an estimate of the relation of this oscillation to desired temperature, θ_d.

In cases where the amplitude of oscillation is too great for the given accuracy specification, suggestions are made for the inclusion of a phase-advance element in the control loop with the aim of reducing the amplitude of oscillation.

Although an on–off temperature control loop may be simple in terms of hardware, it is a nonlinear loop and as such its rigorous analysis is quite difficult and approximate methods will usually be used. The chief attraction of on–off control is the usually low cost of the associated actuators.

10.14 References and further reading

A number of useful techniques for the analysis of nonlinear control systems will be found in:

LEIGH, J. R. (1983): 'Essentials of nonlinear control theory' (Peter Peregrinus)

In particular, the book covers the describing function technique which can be very useful for yielding qualitative explanations of nonlinear system behaviour.

A useful text covering on–off control systems is:

TSYPKIN, Y. Z. (1984): 'Relay control systems' (Cambridge University Press)

Chapter 11

Temperature control – 2: Criteria and techniques for design of controllers for use with continuously variable actuators

11.1 Introduction

In general, a closed loop system is required to meet a specification that is made up of three parts:

(*a*) A speed of response requirement
(*b*) An accuracy requirement
(*c*) A stability margin requirement.

Speed of response is often conveniently characterised by *closed-loop bandwidth* ω_b and the nature of the transient response by the parameter

$$M = \max_{\omega}|(y(\omega)/v(\omega))|$$

For a second-order system, the undamped natural frequency ω_n and the damping factor ζ contain similar information.

The accuracy requirement is usually stated in terms of *steady state error* in response to a particular input such as a step input, ramp input or parabolic input. Relative stability is usually measured in terms of *gain margin* and *phase margin*. The gain margin is the attenuation (dB) when the open-loop phase angle is $-180°$. The phase margin is $(180 - \phi)°$, where ϕ is the phase angle (deg) when the open-loop gain is 0 dB. For linear systems the following main design techniques are available:

(i) Ziegler–Nichols technique
(ii) Synthesis of a system with a known transfer function
(iii) Root-locus design technique
(iv) Bode-diagram design technique
(v) Nyquist-diagram design technique

(vi) Nichols-chart design technique
(vii) Synthesis of an optimal controller such that a given performance index is maximised

All of these techniques will now be illustrated by simple examples. (Techniques (iv), (v) and (vi) display the same frequency response information in different forms, so only one, the Bode design technique, will be illustrated.)

11.1.1 Ziegler–Nichols design approach
In the Ziegler–Nichols approach, every process is approximated by the same form of transfer function

$$G(s) = \frac{Ke^{-ST_d}}{(1 + ST_1)} \tag{11.1}$$

The coefficients K, T_d, T_1 are found by simple graphical construction from the open-loop unit step response of the process as shown in Fig. 11.1.

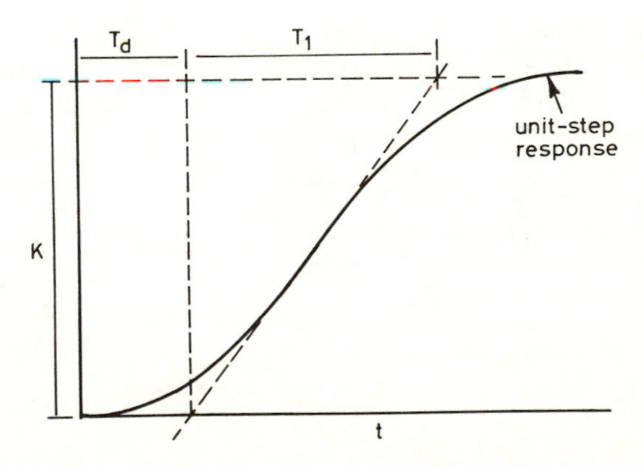

Fig. 11.1 *Illustration of how the coefficients in eqn. 11.1 may be picked off a step-response curve*

Ziegler–Nichols rules then suggest the following alternative controllers:

$$\text{Proportional only, } D_1(s) = C \tag{11.2}$$

$$\text{Proportional + Integral } D_2(s) = C\left(1 + \frac{1}{sT_I}\right) \tag{11.3}$$

$$\text{Proportional + Derivative + Integral } D_3(s) =$$

$$= C\left(1 + \frac{1}{sT_I} + sT_D\right) \tag{11.4}$$

The numerical coefficients that are recommended for the controllers are given in Table 11.1. The settings were derived by Ziegler and Nichols to minimise the integral of absolute error after the application of a step change in desired value. See Leigh (1982) for further background.

Table 11.1 *Ziegler–Nichols coefficients*

	For changes in desired value		
		T_I	T_D
Proportional controller	$\dfrac{T_1}{KT_d}$	–	–
Proportional plus integral controller	$\dfrac{0{\cdot}9T_1}{KT_d}$	$3{\cdot}3T_d$	–
Proportional plus integral plus derivative controller	$\dfrac{1{\cdot}2T_1}{KT_d}$	$2T_d$	$0{\cdot}5T_d$

11.1.2 Illustrative example

An electrically heated oven has the open-loop step response shown in Fig. 11.2. By graphical construction as shown in the Figure we obtain the approximate transfer function

$$G(s) = \frac{60e^{-135s}}{(1 + 285s)} \tag{11.5}$$

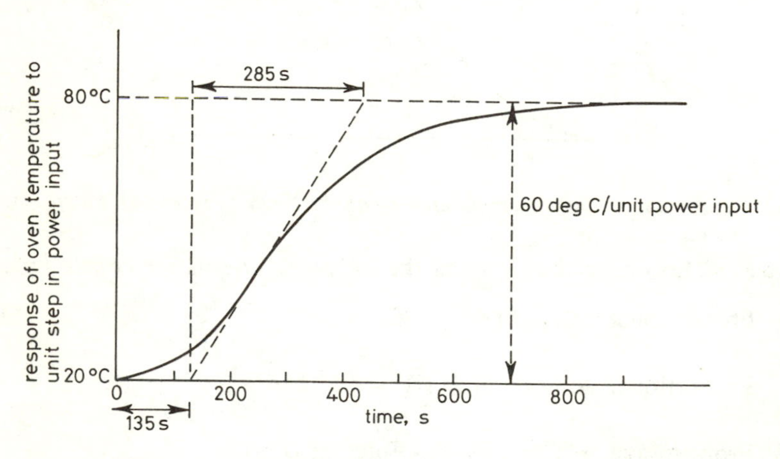

Fig. 11.2 *Step response of an electrical oven*

and applying the Ziegler–Nichols rules leads to the proportional-plus-integral-plus-derivative controller with transfer function

$$D(s) = \frac{(1{\cdot}2)(285)}{(60)(135)}\left(1 + \frac{(135)s}{2} + \frac{1}{(2)(135)s}\right)$$

$$= 0{\cdot}042\left(1 + 67{\cdot}5s + \frac{1}{270s}\right) \tag{11.6}$$

Note that the Ziegler–Nichols rules give no guidance on whether to use a 1-term, 2-term or 3-term controller. However, best performance will usually be obtained by using all three terms.

Note the simplicity of the Ziegler–Nichols approach. Both process modelling and controller design are achieved very rapidly. The result is a robust control loop that over the years has been found to give good control in imprecise industrial situations where all parameters are prone to drift and all signals are, to a greater or lesser extent, corrupted by noise. Note also that no system specification is either needed or is able to be incorporated into the design procedure.

11.2 Synthesis of a system with a known closed-loop transfer function

Given a process of open-loop transfer function $G(s)$ and given a desired closed-loop transfer function $H(s)$, we may write

$$H(s) = \frac{G(s)D(s)}{1 + G(s)D(s)}$$

where $D(s)$ is a controller that precedes the process in the loop (Fig. 11.3).

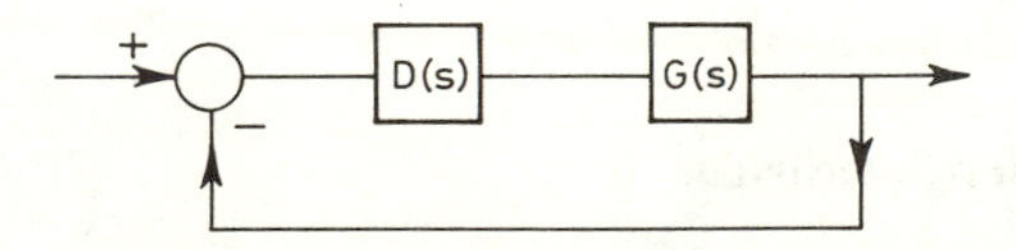

Fig. 11.3 *Controller D(s) designed for control of the process whose model is G(s)*

The controller transfer function $D(s)$ is given by

$$D(s) = \frac{H(s)}{G(s)(1 - H(s))} \tag{11.7}$$

Clearly the desired closed-loop transfer function H must be chosen so that $D(s)$ is a realisable transfer function.

11.2.1 Illustrative example

A process has the open-loop transfer function

$$G(s) = \frac{1}{(1 + 10s)(1 + s)} \tag{11.8}$$

and is required under closed-loop control to have the transfer function

$$H(s) = \frac{1}{(s+1+j)(s+1-j)} = \frac{1}{s^2+2s+2}$$

using the equation

$$D(s) = \frac{H(s)}{G(s)(1-H(s))}$$

we obtain

$$D(s) = \frac{1/(s^2+2s+2)}{\dfrac{1}{(1+10s)(1+s)}(1-1/(s^2+2s+2))}$$

$$= \frac{1}{\dfrac{1}{(1+10s)(1+s)}(s^2+2s+2-1)}$$

$$= \frac{(1+10s)(1+s)}{s^2+2s+1} = \left(\frac{1+10s}{1+s}\right) \tag{11.9}$$

It can be seen that in essence the design technique consists in cancelling process poles and zeros and replacing them by poles and zeros of the controller. Since the controller contains the inverse of the process transfer function, complex processes will always need complex controllers if this method is used. Also a process with right-half-plane zeros will require an unstable controller with right-half-plane poles.

11.3 Root-locus design technique

Instead of cancelling process poles and replacing them with controller poles as in the previous Section, root-locus design attempts to move process poles to new desired locations by the choice of loop gain. Dynamic elements are introduced into the controller only as necessary to obtain the desired closed loop poles.

Let the process transfer function be $G(s) = P(s)/Q(s)$ and let the first attempt at a control design be to set $D(s) = k$, a simple proportional controller. Then $H(s)$, the resultant closed loop transfer function, is given by

$$H(s) = \frac{C(s)D(s)}{1+G(s)D(s)} = \frac{\dfrac{P(s)}{Q(s)}k}{1+\dfrac{P(s)}{Q(s)}k} = \frac{P(s)k}{Q(s)+P(s)k} \tag{11.10}$$

The poles of $H(s)$ lie on the locus in the complex plane that satisfies $Q(s) + P(s)k = 0$ and the location along the locus is a function of gain k. Design consists in choosing k such that the poles of $H(s)$ lie in desired locations. Should no choice of k locate the poles suitably then a dynamic transfer function

$$kD(s) = k\frac{P(s)'}{Q(s)'}$$

is needed for the controller. $H(s)$ is then given by

$$\frac{G(s)kD(s)}{1 + G(s)kD(s)} = \frac{\frac{P(s)}{Q(s)}k\frac{P(s)'}{Q'(s)}}{1 + \frac{P(s)}{Q(s)}k\frac{P'(s)}{Q'(s)}} = \frac{kP(s)P'(s)}{Q(s)Q'(s) + kP(s)P'(s)} \tag{11.11}$$

Design consists in selecting a suitable controller transfer function $P(s)'/Q(s)'$ and then selecting the gain k so that the closed-loop poles lie in the desired locations. As in all interactive graphical methods, a certain subjective skill is needed that can only be obtained with practice.

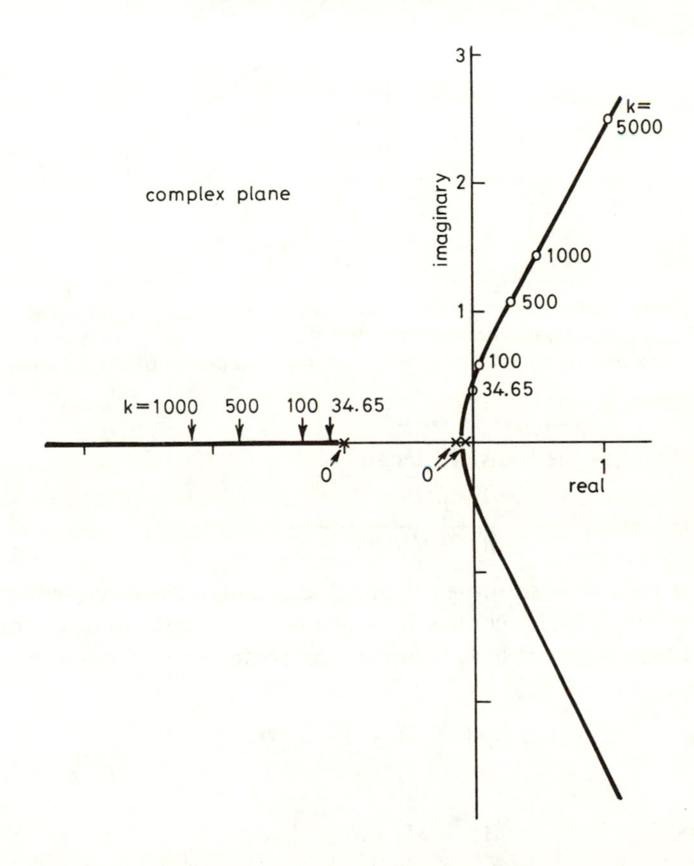

Fig. 11.4a *Root locus for the case where the process described by eqn. 11.12 is under closed-loop control by a proportional-only controller of gain k*
k = gain of controller

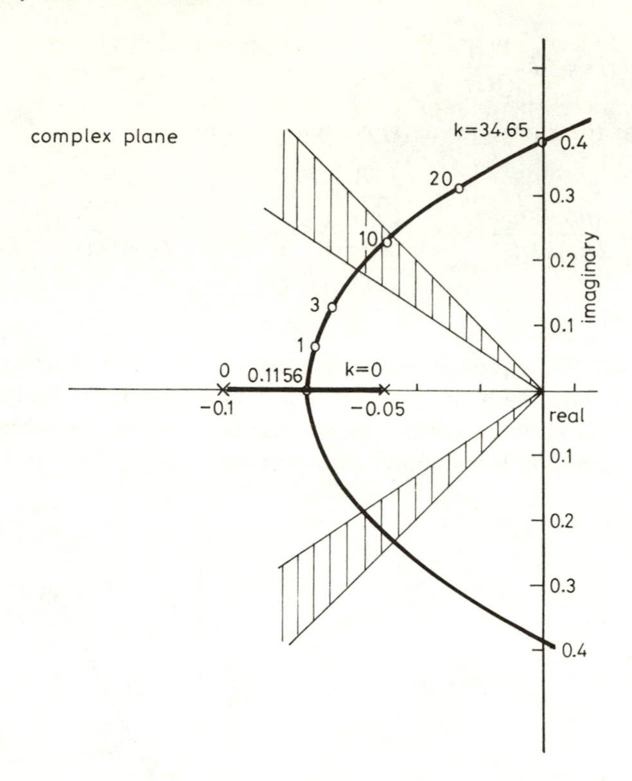

Fig. 11.4b *Root locus (enlarged scale): note unequal horizontal/vertical scales*
Gain k must be chosen such that the closed-loop poles lie within the shaded area (see Section 11.3.1 for further information on choice of shaded area)

11.3.1 Illustrative example

A process $G(s)$ has the transfer function

$$G(s) = \frac{1}{(1 + 20s)(1 + 10s)(1 + s)} \tag{11.12}$$

We plot the root locus showing the possible locations of the closed-loop poles when the system is under control by a proportional-only controller of gain k. The closed-loop poles are at locations in the complex plane given by solutions of the equation

$$(1 + 20s)(1 + 10s)(1 + s) + k = 0$$

or

$$200s^3 + 230s^2 + 31s + (1 + k) = 0$$

The root locus is plotted in Fig. 11.4 with gain k as a parameter. A suitable choice of k for this example would be $k = 10$ to ensure that poles fall within the shaded region in Fig. 11.4. A system with poles in the shaded region will have a transient response that is a reasonable compromise between speed of response and sufficient stability margin. Improved control may be obtained by introducing dynamic elements into the controller. Such a strategy allows the root locus to be shaped to that it will pass through any given region in the complex plane.

11.4 Bode design techniques

The Bode diagram consists of two curves representing frequency-response information: a plot of magnitude (dB) against logarithm of frequency, and a plot of phase angle (degrees) against logarithm of frequency. The gain margin and phase margin of a system may easily be read off the Bode diagram. The technique is particularly useful for systems with dead time since this affects only the phase plot but not the gain plot in the Bode diagram.

11.4.1 Illustration of the use of the Bode diagram as a control design tool in an application with a measurement delay

Fig. 11.5 shows a simple control problem. A constant flow of fluid passes over

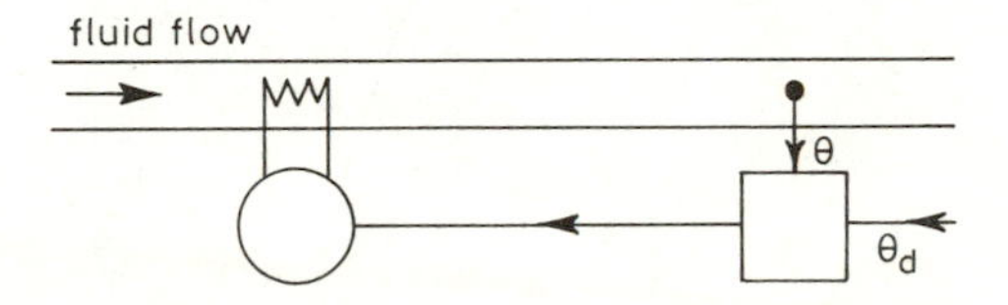

Fig. 11.5 *Fluid-temperature control problem*

a heating element. The aim is to control the measured temperature to be equal to a given desired temperature θ_d. A proportional-plus-integral controller is to be used. The Bode diagram will be used to design a controller.

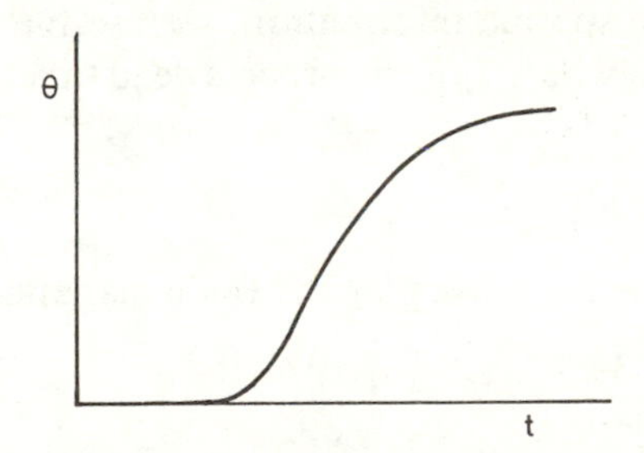

Fig. 11.6 *Open-loop step response for the system of Fig. 11.5*

11.4.2 Procedure

Assume that open-loop step tests have produced step-response curves of the form given in Fig. 11.6, and that the process transfer function $G(s) = \theta(s)/u(s)$

has been approximated by the expression

$$G(s) = \frac{e^{-10s}}{(1 + 2s)} \tag{11.13}$$

The frequency response of $G(s)$ is plotted in the Bode diagram (Fig. 11.7). The proportional-plus-integral controller has the transfer function

$$D(s) = C\left(1 + \frac{1}{sT_I}\right)$$

with C, T_I being coefficients to be chosen:

$$D(s) = C\left(\frac{sT_I + 1}{sT_I}\right) \tag{11.14}$$

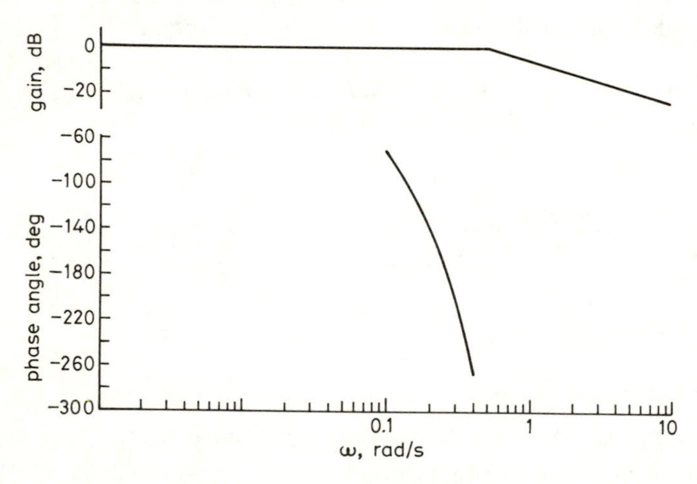

Fig. 11.7 *Frequency response of the process $e^{-10s}/(1 + 2s)$*

In the absence of more specific information, we use the approach of Section 11.1.1 and set $T_I = 3T_d$, where T_d is the process dead time, so that $D(s)$ becomes

$$D(s) = C\left(\frac{1 + 30s}{30s}\right) \tag{11.15}$$

Leaving the gain C to be fixed, we plot the Bode diagram for the combination

$$\frac{e^{-10s}}{(1 + 2s)} \frac{(1 + 30s)}{30s}$$

in Fig. 11.8.

A well established design technique is to choose the gain C so that a gain margin of at least 8 dB and a phase margin of at least 30° are obtained. From the Figure, it can be seen that both criteria will be satisfied if attenuation equivalent to 8 dB is introduced into the control loop; i.e. we require $C = 0{\cdot}4$ to yield a controller of the form

$$D(s) = 0{\cdot}4\left(1 + \frac{1}{30s}\right)$$

If this design exercise were part of a real application, it would be essential to consider the effect of a possible reduction in the flow rate on the stability. To ensure stability at all flow rates, the loop gain would need to be kept to no more than (say) −8 dB at all frequencies. Such a requirement could only be met by removal of the integrating term from the controller, with consequent loss of its error correcting facility.

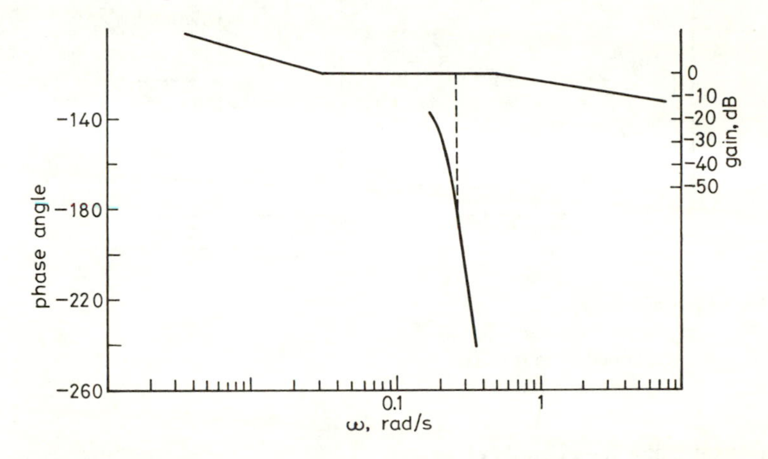

Fig. 11.8 *Frequency response of the combination $e^{-10s}(1 + 30s)/((1 + 2s)30s)$*

The exercise illustrates well the difficulties caused to the control engineer by the presence of the long dead time in the process. In terms of the Bode diagram, the effect of the long dead time is shown in the very steep slope of the phase characteristic. A first reaction might be to attempt to improve the situation by providing phase advance by the use of a derivative term in the controller. However, a process with dead time cannot benefit from such an addition. To understand why this is so, we refer to the two curves sketched in Fig. 11.9, showing, respectively, the step response of a high-order process and of a dead time process.

In the case of the high-order process, the derivative of the response (shown dotted) gives early information on the response that is to come, and this information can be used to advantage for control purposes. The derivative of the dead-time process, in contrast, gives no advance information on the response that is to come, and therefore the addition of a derivative term cannot improve the control of a dead-time process.

Improved controllability for a situation such as that of Figure 11.5 should first be sought from physical modifications to the process to attempt a reduction in the measurement dead time. Alternatively, mathematical compensation for the dead time may be achieved by the use of specialist control algorithms that are designed for that purpose. Such algorithms rely, in principle, on the use of mathematical models to subtract out the dead time from the control loop. Given a perfect process model, complete compensation for the dead time may be achieved. In practice, schemes making use of dead-time compensation algorithms need to be designed with due regard to the envelope of changes that can occur in the process and the inevitable mismatch between a process and its supposed model. Methods for compensation of dead time will be described in Section 11.8.

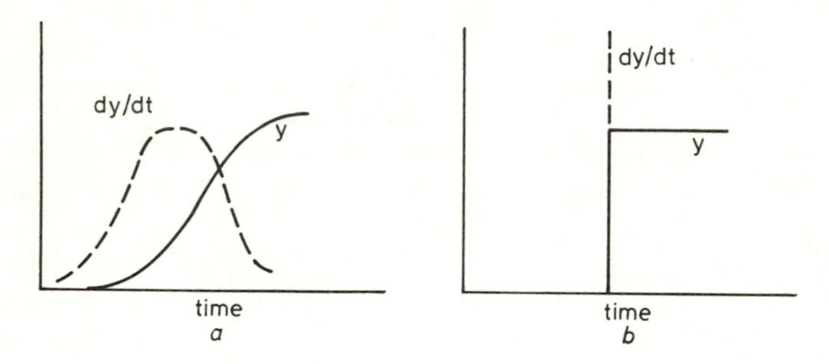

Fig. 11.9 *Step response*
a High-order process
b Purely dead-time process

11.5 Design using the Nichols chart

The frequency-response information that is contained in a Bode plot may be represented with equal validity in a single plot on axes of gain (dB) against phase angle (degrees). In this case, frequency is a parameter along the response curve.

Gain and phase margins may be read off from the gain/phase plot just as easily as from the Bode diagram. In addition, provided that the loop is to be closed directly through unity feedback, closed-loop performance information may be read off from the gain/phase plot. This is achieved by superimposing a Nichols-chart overlay on to the gain/phase plot. The Nichols chart consists of families of loci of constant M and ϕ for the unity-feedback closed-loop system where

$$M = 20 \log_{10} \frac{|y(j\omega)|}{|v(j\omega)|} \tag{11.16}$$

$$\phi = \angle v(j\omega) - \angle y(j\omega) \tag{11.17}$$

Thus the open-loop gain/phase plot, together with its Nichols-chart overlay, can yield stability-margin information and valuable information on the response of

Table 11.2 *Magnitude and phase angle against frequency for the transfer function $1/s(s + 1)$*

ω (rad/s)	Magnitude (dB) $20 \log (1/\omega(1 + \omega^2)^{1/2})$	Angle (deg) $-90 - \tan^{-1}\omega$
0·1	20	−96
0·25	11·77	−104
0·4	7·3	−112
0·5	5	−116
1	−3	−135
1.5	−8·6	−146
2	−13	−153
3	−19·5	−162

the closed-loop system to sinusoidal inputs, allowing resonant frequency, maximum magnification, closed-loop bandwidth and closed-loop phase angle to be determined.

The Nichols chart contains exactly the same information as does the Bode diagram, and it can be used as an alternative to that diagram as a tool for analysis and design. However the Nichols chart has a useful additional feature by the use of its overlay; the Nichols chart allows salient parameters of the closed-loop performance to be read off from the open-loop curve – *provided that there is unity feedback; i.e. that there are no components in the feedback loop.* Further, the Nichols chart affords a simple graphical technique for choosing suitable loop gains. To some extent, this graphical facility has been rendered obsolescent by computer programs that can easily calculate closed-loop parameters from open-loop equations; however, the technique remains valuable in aiding 'feel' and interactive design.

A very simple illustrative example is given. A heating process of transfer function $G(s) = 1/s$ is in series with a controller of transfer function $D(s) = k/(1 + s)$. Use the Nichols chart to fix the value of controller gain k so that, in unity-feedback closed loop, the maximum amplification of any incoming sinusoidal frequency shall be 3 dB. We take the transfer function $G(s)D(s)$ of the combined controller and plant and, replacing s by $j\omega$ and setting (temporarily) k to unity, we obtain a complex number whose magnitude (dB) we plot against the corresponding angle (degrees) to form, in conjunction with the overlays, the Nichols chart. The calculation of magnitude and phase angle is given in Table 11.2.

These data are plotted in Fig. 11.10 to produce the locus shown. We now have to decide: by how much must this locus be slid vertically upwards in order that it shall become tangential to the 3 dB curve on the overlay. This operation is performed most easily using a movable transparent overlay, but, since this is difficult to illustrate, we work by inspection and argue that an upward movement of 5 dB will produce this tangential condition. Clearly, the upward movement of 5 dB can be obtained physically by setting the controller gain k to k = antilog (5/20) = 1·8. The Nichols chart also shows that the frequency where the maximum magnification will be obtained is approximately ω = 1 rad/s.

The Nichols chart is also useful where specific noise-rejection characteristics have to be designed into a system. In such a case, the system transfer function is first manipulated into the form $y(s)/w(s)$, where y is the system output and w is the disturbance input. The analysis then proceeds exactly analogously to that shown in the example.

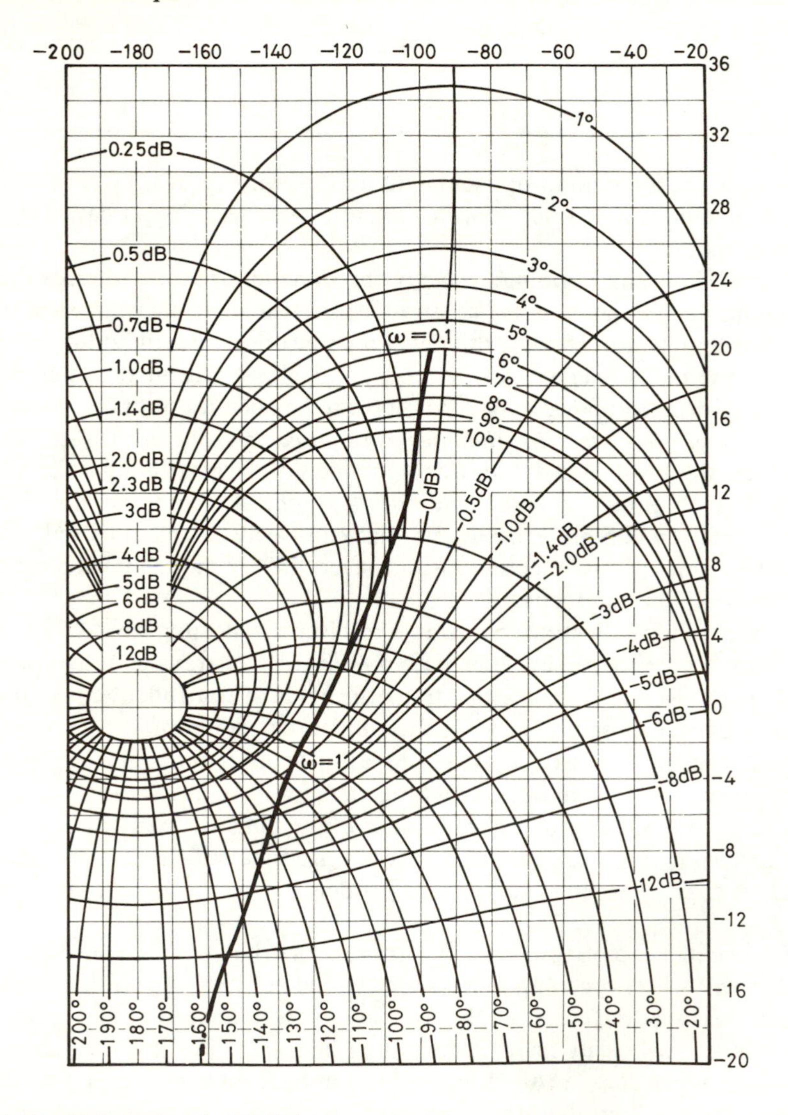

Fig. 11.10 *Nichols chart for k/s(1 + s)*

11.6 Synthesis of an optimal controller such that a given performance index is maximised

Optimal control techniques may be used to determine the best possible strategy (as measured by a performance index) for a particular situation; e.g. the time/temperature curve that will heat-treat an alloy in minimum time, or the minimum-cost firing strategy for a plant using multiple fuels, can be generated by optimum control techniques.

Here we demonstrate optimum control by means of a very simple example; see Noton (1965) for numerical details. Refer to Fig. 11.11 and suppose that we

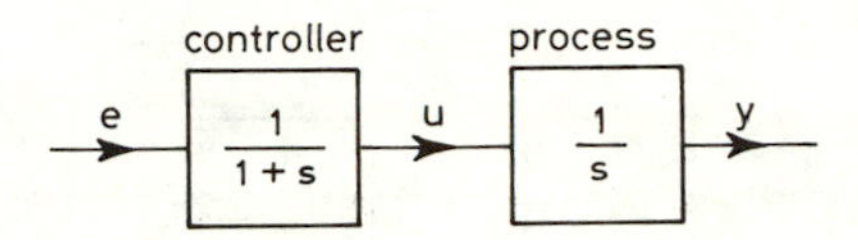

Fig. 11.11 *A simple optimisation problem*

require the value of the output y to be kept always close to zero despite the effect of unpredictable disturbances. This requirement might be turned into a mathematical statement that the cost function

$$\int_{t_0}^{t_1} y^2 \, dt, \qquad t_1 > t_0$$

has to be minimised. However, without further modification, the problem is not well posed since the solution will be to use an unbounded value of e to bring y to zero in infinitely short time whenever it is perturbed. To progress, either e must be *constrained*, by imposing limits on the range of its magnitude, or else e must be included in the cost function in such a way that a cost is incurred whenever e is non-zero. Here we choose the second approach by selecting the modified cost function

$$J = \int_{t_0}^{t_1} (y^2 + ae^2) \, dt, \qquad t_1 > t_0$$

in which the value of the constant a determines the weight to be associated with the quantity e, compared with the unity weight that is attached to the quantity y. Putting $a = 0{\cdot}1$, $t_0 = 0$, $t_1 = \infty$ and solving using any one of the standard optimisation approaches (calculus of variations, Pontryagin's maximum method or dynamic programming) leads, after considerable manipulation, to the following equations for u and e *as functions of time*:

$$u = Ae^{-1{\cdot}35t} \sin(1{\cdot}15t + B) \qquad (11.18)$$

$$e = Ae^{-1{\cdot}35t}(1{\cdot}15 \cos(1{\cdot}15t + B) - 1{\cdot}35 \sin(1{\cdot}15t + B)) \qquad (11.19)$$

in which A and B are arbitrary constants to be determined.

Note that the solution here is typical. Optimisation techniques do not naturally produce feedback laws. However, because here $t_1 = \infty$, the controls to be applied will be linear combinations of the system states and a feedback law can be obtained by substitution. Here we have

$$\left.\begin{aligned} A \sin B &= y(0) \\ A \cos B &= 0{\cdot}867u(0) + 1{\cdot}173y(0) \end{aligned}\right\} \quad (11.20)$$

and because time now may always be considered as $t = 0$ (since the upper limit of integration is always infinitely far away), then we finally obtain as the optimal feedback control law

$$e = -1{\cdot}706y - 3{\cdot}162u \quad \text{(see Fig. 11.12)} \qquad (11.21)$$

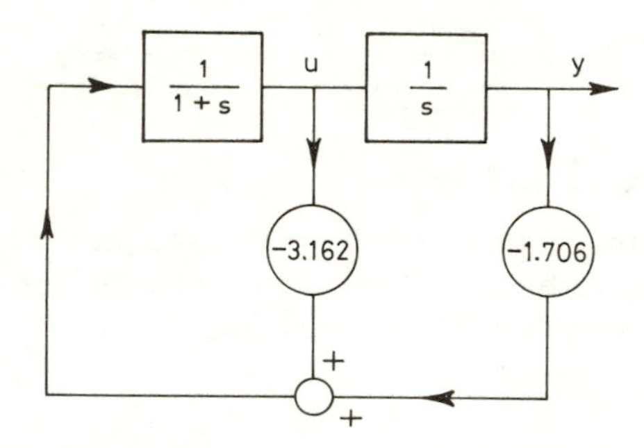

Fig. 11.12 *Optimal feedback solution*

Optimisation techniques have great potential in temperature-control applications where they may allow minimum-time heating strategies or minimum-fuel-consumption policies to be calculated. However, optimisation techniques are relatively complex, and there is a considerably tendency for problems to be simplified to the point of near meaninglessness to allow the techniques to be applied, and this tendency must be guarded against. There is a vast literature on optimisation from which only three representatives have been quoted in Section 11.9.

11.7 Alternative approaches to obtain pseudo-continuous control of an electrical heater

Where an electrical resistance heater is to be driven by a PID controller, or by any controller that requires a continuously variable electrical power to be produced, a difficulty arises: no reasonably low-cost device exists that will provide a truly continuously variable power output. However, two methods are available to provide pseudo-continuous variation of power level. These are:

(*a*) 'Chopping' of sinusoids
(*b*) Pulse-width modulation.

(*a*) *Chopping of sinusoids*

A common method of controlling the mean level of an alternating voltage is to vary the firing angle of a controlled rectifier as shown in Fig. 11.13, so that some initial segment of each sinusoid is held back and hence any desired mean voltage level may be obtained. Such arrangements are used very satisfactorily in temperature-control applications. The main disadvantage of the arrangement is its relatively high cost compared with the pulse-width-modulation strategy to be described in (*b*).

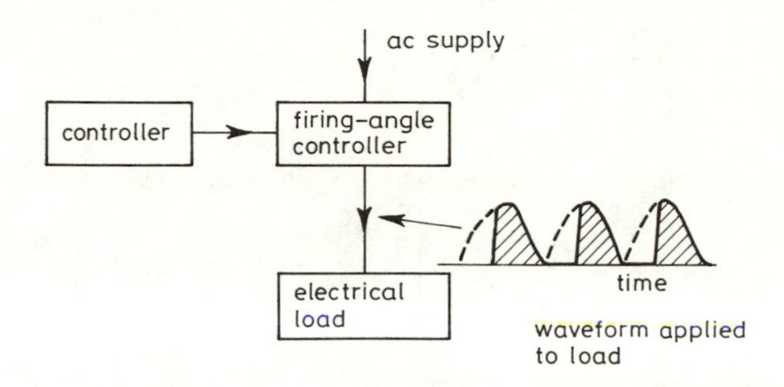

Fig. 11.13 *'Chopping' of sinusoids*

(*b*) *Pulse-width modulation to obtain pseudo-continuous control from an on–off actuator*

Since on–off actuators are usually significantly cheaper than their continuously variable equivalents, it is worth considering their use even in those cases that seem to call for continuous actuation.

The basic approach is to control the ratio of 'time on' to 'time off' so as to produce a (pulse-width-modulated) rectangular wave whose mean level is equal to that required (Fig. 11.14). The frequency of the rectangular wave must be well

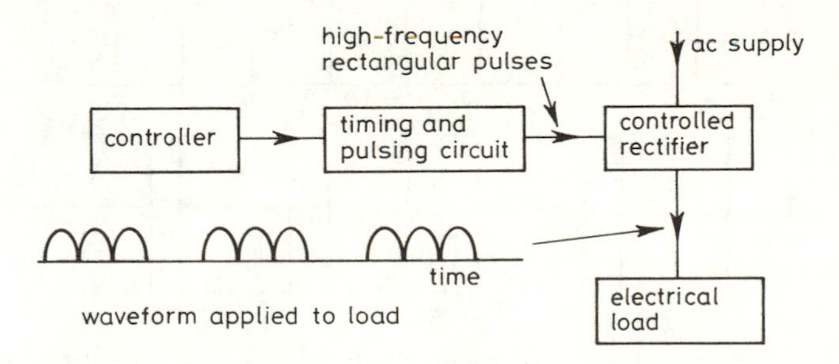

Fig. 11.14 *Pulse-width modulation*

above the closed-loop bandwidth of the control system to ensure that the process is affected only by the mean level of the rectangular wave, and there is no 'following' of the rectangular wave by the process. To be more specific, let ω_b be the closed-loop bandwidth of the control system and let T_R be the period of the rectangular wave. Fourier decomposition of the rectangular wave shows that the lowest frequency present is $1/T_R$ Hz, and thus we require that

$$\frac{1}{T_R} \gg \omega_b,$$

in order for apparently continuous control to be obtained, as explained in Fig. 11.15*a* and *b*).

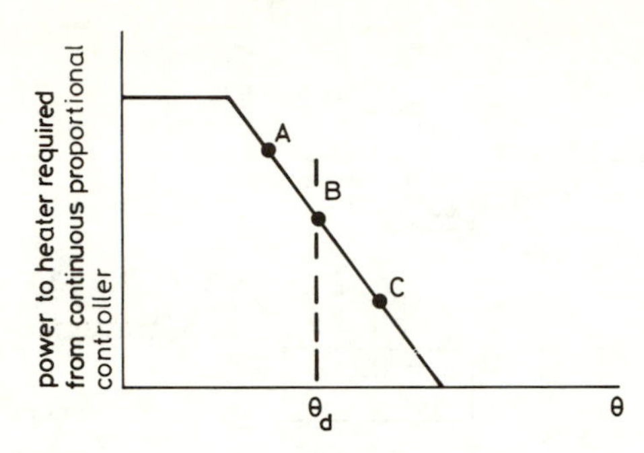

Fig. 11.15a *Continuous controller characteristic that is to be approximated*

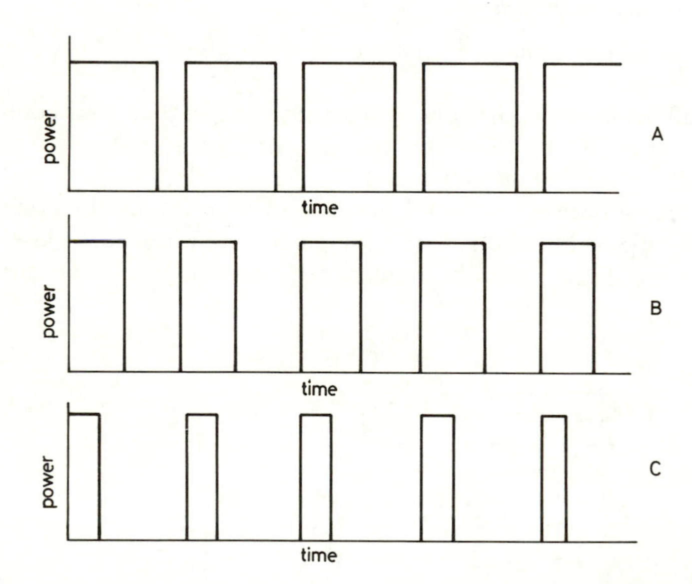

Fig. 11.15b *Corresponding on–off behaviour of the pulse-width-modulated controller*

Although only one parameter (on-time/off-time ratio) can be manipulated, it is nevertheless easy to arrange for 3-term control action to be obtained. Fig. 11.16 illustrates, without the need for supporting text, how proportional and proportional-and-integral actions are achieved by pulse-width modulation.

Controllers for electrical heating use a stream of high-frequency pulses to open a silicon-controlled rectifier switch during the on period; an absence of pulses produces the off period. There is no need for a circuit to control the firing angle of the silicon-controlled rectifier as is required in the alternative schemes that 'chop' each power wave to obtain another, closer, approximation to continuous control.

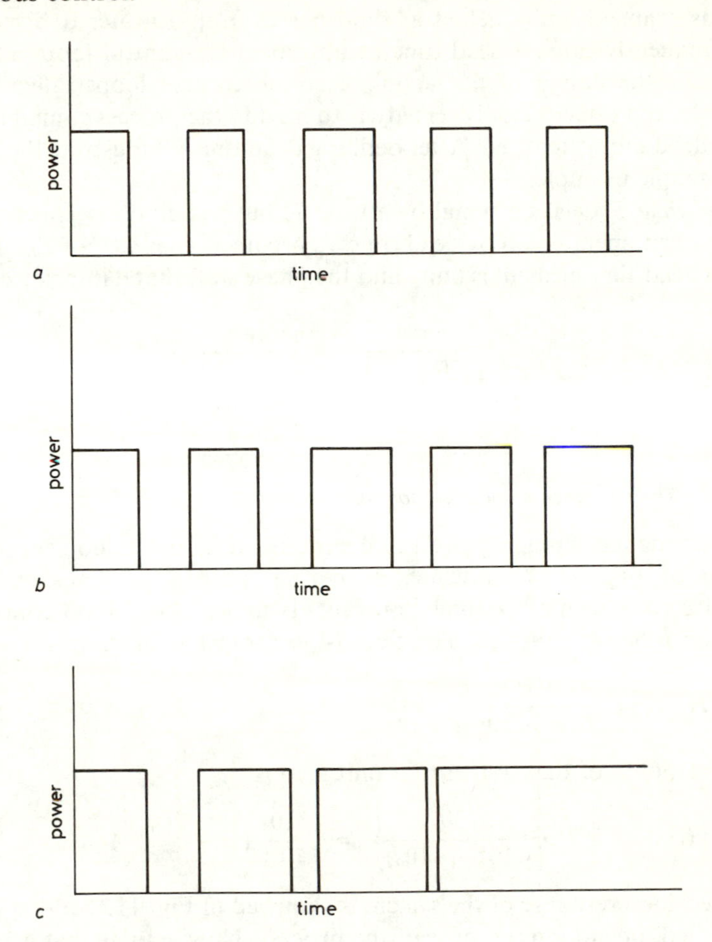

Fig. 11.16 *How proportional action is achieved*
a Proportional control only
b Proportional plus weak integral action
c Proportional plus strong integral action
Graphs show open-loop behaviour of controller

11.8 Control of processes with dead time

We already encountered dead time in passing in Section 11.4. Here we consider more specifically the degrading effect of dead time on control-system performance.

Controlling the temperature of liquid in a well stirred tank is straightforward because the dynamics of the process are first order and a high loop gain can therefore be implemented without danger of instability. However, many temperature-control problems involve fluids travelling along pipes, and, in such situations, transportation delays as fluid passes from a heater to a measuring sensor frequently cause a dead time to appear in the control loop. Such dead times make the design of fast-acting accurate control loops difficult if not impossible, and often the only remedy is to modify the process configuration so that the dead time is reduced. After defining dead time we illustrate the problem using a simple example.

A *dead time* T_d delays a signal by a time T_d but has no other effect. A signal $a \sin \omega t$ when operated on by dead time T_d becomes $a \sin \omega(t - T_d)$. Thus the gain of a dead-time element is unity and the phase shift that it introduces is ωT_d.

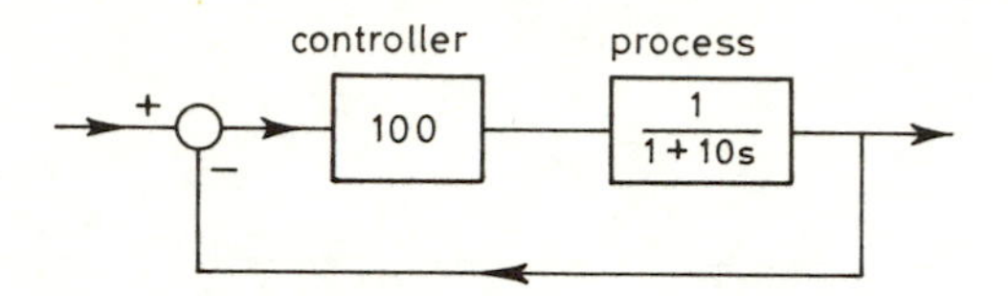

Fig. 11.17 *Thermal process under proportional control*

To illustrate the degrading effect of dead time in a control loop consider first the loop of Fig. 11.17 in which a thermal process of transfer function $1/(1 + 10s)$ (i.e. with a 10 s time constant) is under closed-loop control by a simple controller of gain 100. The closed-loop transfer function is

$$H(s) = \frac{100}{101 + 10s} \tag{11.22}$$

and the response of this system to a unit step is

$$y(t) = \mathscr{L}^{-1}\left\{\frac{100}{s(101 + 10s)}\right\} = \frac{10}{10{\cdot}1}(1 - e^{-10{\cdot}1t}) \tag{11.23}$$

The closed-loop response of the system is sketched in Fig. 11.18 along with the uncontrolled, open-loop response of the process. Now assume that a 10 s dead time appears in the process. We investigate the closed-loop control of this augmented process (Fig. 11.9). We could investigate graphically using the Bode diagram as in Section 11.4..

For the moment we simply ask: at which frequency ω^* will the phase shift in the process be $-180°$. We know that the phase shift in the process is given by

$$\phi = -\omega T_d - \tan^{-1} 10\omega \tag{11.24}$$

and we therefore seek ω to satisfy

$$\pi = 10\omega + \tan^{-1} 10\omega \tag{11.25}$$

Trial and error using a calculator quickly yields

$$\omega^* = 0{\cdot}2029 \text{ rad/s} \tag{11.26}$$

The gain of the process at frequency ω^* is given by

$$\text{process gain} = \frac{1}{(1^2 + (10\omega^*)^2)^{1/2}} = 0{\cdot}44 \tag{11.27}$$

so that setting the controller gain k to $k = 1/0{\cdot}44$ would bring the loop to its stability limit.

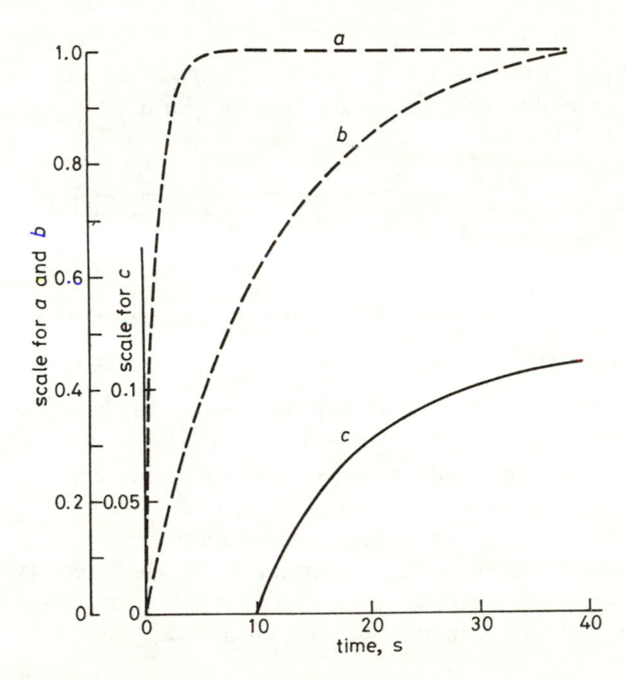

Fig. 11.18 *Comparison of step responses*
Note that unequal vertical scales have been used
a Step response of system without delay in closed loop
b Step response of process in closed loop
c Step response of system including delay in closed loop using maximum allowable controller gain

Using the usual criterion that there should exist a stability margin of 8 dB leads to a recommended controller gain of

$$k = \frac{1}{0{\cdot}44} \times \frac{1}{20 \log_{10} 8} = 0{\cdot}125 \tag{11.28}$$

With this controller implemented, the closed-loop transfer function becomes

$$H(s) = \frac{\dfrac{0{\cdot}125e^{-10s}}{1 + 10s}}{1 + \dfrac{0{\cdot}125e^{-10s}}{1 + 10s}} = \frac{0{\cdot}125e^{-10s}}{1 + 10s + 0{\cdot}125e^{-10s}} \tag{11.29}$$

and the response of the closed-loop system to a unit input step demand is

$$y(t) = \mathscr{L}^{-1}\left\{\frac{0{\cdot}125e^{-10s}}{s(1 + 10s + 0{\cdot}125e^{-10s})}\right\} \quad (11.30)$$

Using the final-value theorem we see at once that the steady-state value of $y(t)$ will be

$$y(t)_{\text{steady state}} = \frac{0{\cdot}125}{1{\cdot}125} = 0{\cdot}111 \quad (11.31)$$

Thus the output of the closed-loop system in response to a demand is now so far from the desired value in the steady state that for most purposes the system would be considered unusable.

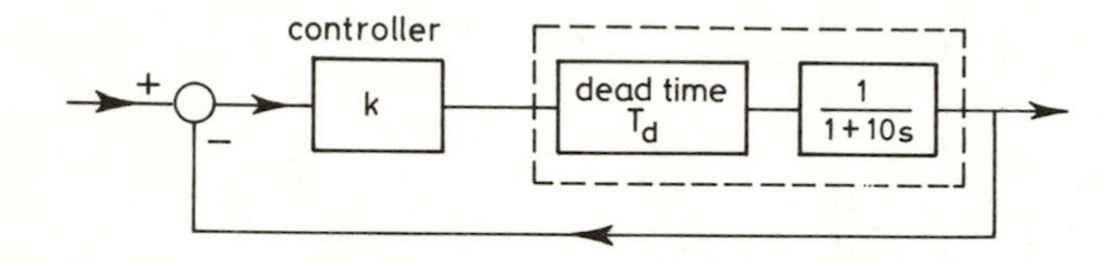

Fig. 11.19 *As Fig. 11.17, but now a dead-time T_d is incorporated into the process*

The complete response of the closed-loop system has been sketched in Fig. 11.18. Recall that this is the best response that can be obtained without moving close to the stability limit and then the magnitude of the degrading effect of the process dead time is readily appreciated. Of course, other terms may be added to the controller to attempt to improve the loop performance, and certain special procedures are available (see Section 11.9). However, unless the dead time can be removed from the control loop, no algorithmic solution will, in practice, usually improve matters very significantly.

11.9 References and further reading

BRYSON, A. E., and HO, Y. C. (1975) 'Applied optimal control' (John Wiley)

EBERHARDT, P., ERLBACHER, J., ERNST, D., and THOMA, M. (1980): 'The control behaviour of temperature controllers based on microprocessors'. Proceedings of the Interkama Congress 'Measurement and automation techniques', Düsseldorf and Berlin (Springer–Verlag) pp. 561–578

FUHRT, B. P., CARAPIC, M., VAN NAUTA LEMKE, H. R., and VERBRUGGEN, H. B. (1977): 'Comparison of two online algorithms for finishing mill temperature control'. Proceedings of the IFAC Conference 'Digital computer applications to process control', pp. 233–239

GAWTHROP, P. J. (1986): 'Self tuning PID controllers: Algorithms and implementation', *IEEE Trans.*, **AC 31,** pp. 201–209

GREIG, D. M. (1980): 'Optimisation' (Longman, London)

JACOBS, D., and DONAGHEY, L. F. (1977): 'Microcomputer implementation of direct digital control algorithms for thermal process control applications', *Trans. ASME Ser. G, J. Dyn. Syst. Meas. & Control,* **99,** pp. 233–240

LEIGH, J. R. (1982): 'Applied control theory', (Peter Peregrinus)

LEIGH, J. R. (1984): 'Applied digital control: theory, design and implementation', (Prentice-Hall International)
LOBODZINSKI, W., ORZYLOWSKI, M., and RUDOLF, Z. (1980): 'Multichannel temperature controller for diffusion furnace', *Euromicr. J. (Netherlands)*, **6**, pp. 325–329
MORRIS, A. J. (1981): 'Self-tuning control of some pilot plant processes', *Microprocessors & Microsystems* **5**, pp. 3–12
NOTON, A. R. M. (1965): 'Introduction to variational methods in control engineering' (Pergamon, Oxford)
ROWLAND, J. R. (1986): 'Linear control systems, modelling, analysis and design' (John Wiley)
UNBEHAUEN, H., SCHMID, Chr., and BÖTTIGER, F. (1976) 'Comparison and application of DDC algorithms for a heat exchanger', *Automatica,* **12,** pp. 393–402
URONEN, P., and YLINIEMI, L. (1977): 'Experimental comparison and application of different DDC algorithms'. Proceedings of IFAC Conference, 'Digital computer applications to process control', pp. 457–464

Chapter 12

Temperature control – 3: Control of spatial temperature distribution

12.1 Illustrative example

Suppose that a cube of material at 20°C is suddenly immersed into an oven held at 1000°C (Fig. 12.1). As time progresses, the temperature within the cube will increase, and at any particular time there will be a temperature distribution as

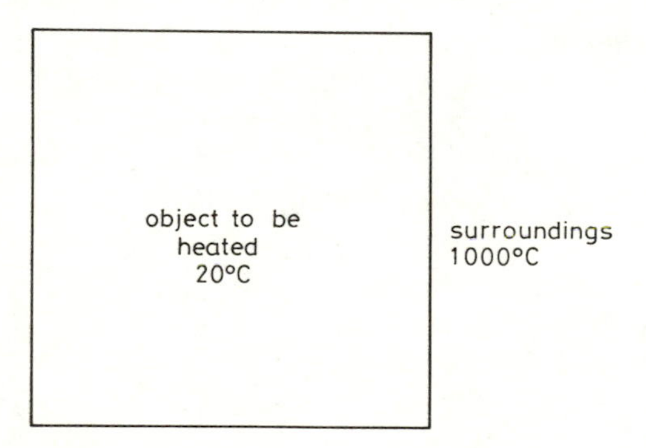

Fig. 12.1 *A solid cube at 20°C is suddenly plunged into a hot environment at $t = 0$*

shown diagrammatically in Fig. 12.2. In many heating processes, it will be the aim to remove the product (corresponding to the cube) once the centre temperature has reached a specified temperature or, alternatively, once the temperature within the product has reached a specified uniformity. Both aims imply a knowledge of internal temperature that cannot be obtained by routine measurement. Calculation of the temperature distribution requires a knowledge of heat-transfer rates that is normally difficult to obtain to a sufficient degree of accuracy.

Of course, if heat can be generated within the cube, for instance by electrical means, the problem is considerably modified and possibly considerably simplified.

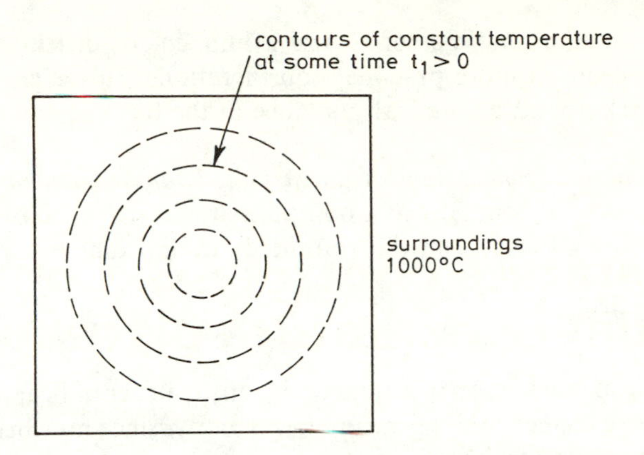

Fig. 12.2 *Illustrating approximately how the temperature distribution in the cube of Fig. 12.1 evolves with time*

12.2 Distributed-parameter models of thermal processes

An assumption that the temperature in a region is uniform is called a lumped-parameter approximation (a single temperature is allocated to represent the whole region). Such an approximation leads on to a model containing ordinary differential equations. Where temperature distributions in space have to be modelled, the dynamic equations are partial differential equations in time and space variables. During a transient, the changes do not take place simultaneously in all parts of the region: rather the effects propagate through the region with a finite velocity.

Let f_{max} represent the highest frequency of interest to which a process is subjected; then provided that the largest physical distance l in the process satisfies

$$l \ll \frac{1}{f_{max}} \tag{12.1}$$

a lumped parameter approximation will be justifiable on theoretical grounds. (Roots, 1969) (Eqn. 12.1 is a statement that, if the physical size of the region to be modelled is much smaller than the shortest wavelength of external stimuli, then the speed of propagation of effects within the region may be regarded as instantaneous.)

As a simple example, suppose that a block of metal is put into an environment where the temperature oscillates sinusoidally with a period of 10 min. What is the largest block that may justifiably be modelled by a lumped-parameter equation? The frequency is 1/600 Hz. Then, from eqn. 12.1,

$$l_{max} \ll \frac{1}{600} \text{ metre}$$

Usually, the decision on whether to use a lumped- or distributed-parameter model will depend on more practical considerations. However, it is useful to have some background theoretical guidance in the form of eqn. 12.1.

12.2.1 Mathematical foundations of a distributed parameter model

Let g be the vector of heat flow at a point p that is surrounded by a volume dv. Then the outflow of heat from the volume dv in unit time is given by

$$\operatorname{div} g\, dv = -\frac{dQ}{dt} \tag{12.2}$$

Considering heat conduction in a rigid body, any transfer of heat dQ must cause a corresponding change in temperature in the body, since no work can be done by expansion. Hence we have

$$dQ = C\varrho\, dv\, d\theta \tag{12.3}$$

where C is the specific heat of the material, of the body ($C = C_p = C_v$ for a rigid body) and ϱ is its density. Combining the results we obtain

$$\operatorname{div} g = -C\varrho \frac{d\theta}{dt} \tag{12.4}$$

Now *Fourier's law* determines the relation between g and θ: For an isotropic medium,

$$g = -\alpha \operatorname{grad} \theta \tag{12.5}$$

i.e. the temperature distribution determines the direction and magnitude of the heat flow at every point in the solid body. α is the *thermal conductivity* of the material.

From the last two equations we obtain

$$-\alpha \operatorname{div} \operatorname{grad} \theta = -c\varrho \frac{d\theta}{dt} \tag{12.6}$$

$$\operatorname{div} \operatorname{grad} \theta = \frac{c\varrho}{\alpha} \frac{d\theta}{dt} \tag{12.7}$$

but

$$\operatorname{div} \operatorname{grad} = \nabla^2 = \frac{\partial^2}{\partial x^2} + \frac{\partial^2}{\partial y^2} + \frac{\partial^2}{\partial z^2}, \tag{12.8}$$

usually known as the Laplace operator. Putting

$$K = \frac{\alpha}{c\varrho}$$

we obtain

$$\nabla^2\theta = \frac{1}{K}\frac{\partial\theta}{\partial t} \tag{12.9}$$

and this is the fundamental differential equation of heat conduction. The solution of the equation has exactly the same form as the expression for the Gaussian probability function, confirming the probabilistic nature of the heat-conduction mechanism. In particular, let a long thin bar of material have a high temperature at one isolated central point at $t = t_0$. The solution of eqn. 12.9 describes how the temperature distribution for $t > t_0$ will be described by ever flatter bell-shaped Gaussian curves as t increases.

12.2.2 What useful guidelines can be formulated from the solution of eqn. 12.9?

A solid material initially at uniform temperature $\theta = 0$ occupies a semi-infinite region $x > 0$. The surface at $x = 0$ is suddenly exposed to a higher temperature $\theta = \theta_0$. What can we say about the penetration of the increased temperature into the interior of the material? Fig. 12.3 shows how the temperature distribution changes with time.

A full solution of eqn. 12.9 shows that any particular temperature, say $\theta_0/2$, (see Fig. 12.3) penetrates into the interior a distance x that is proportional to $t^{1/2}$ – this behaviour is characteristic of diffusion equations in general.

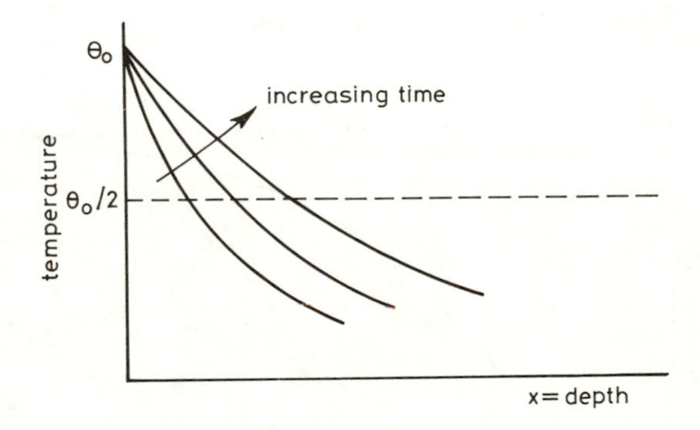

Fig. 12.3 *Penetration of temperature increase into a solid as a function of time*

12.3 Controlling temperature distribution in solids

Many industrial processes require batches of solid objects to be heated to produce a particular temperature distribution. Often the requirement takes the form that the centre temperature in the solid must reach a specified minimum temperature while the surface of the solid must at no time exceed a specified maximum temperature. Examples are: heating of metal blocks for forging, heat treatment of metals and the cooking of foods.

Temperature control consists in varying the boundary conditions that govern the pattern of heat transfer into the solid body until it is calculated that the required temperature distribution has been reached. Control is achieved, not by feedback, but by maintaining a pre-calculated temperature/time profile at the boundary of the solid object. Figs. 12.4 and 12.5 illustrate the principles.

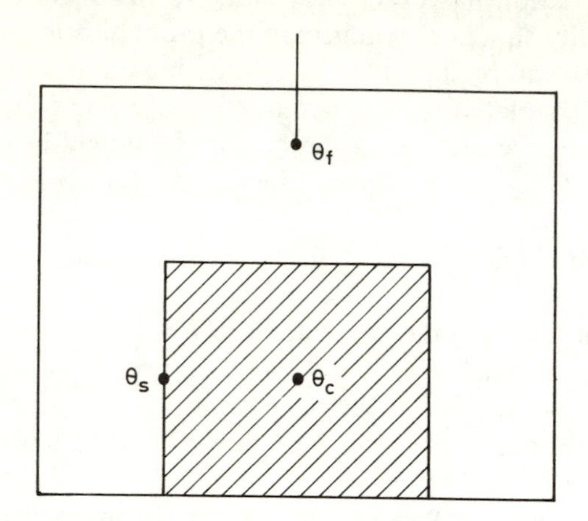

Fig. 12.4 *Heating of a solid body*
θ_f = furnace temperature
θ_s = unmeasured surface temperature in solid block
θ_c = unmeasured centre temperature in solid block

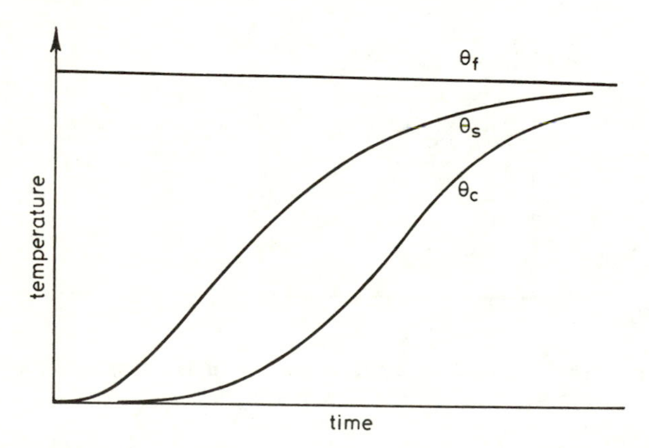

Fig. 12.5 *Heating of a solid body*
Only θ_f can be controlled directly. θ_s and θ_c approach θ_f asymptotically as shown

The calculation of the temperature distribution in a heated solid can be a very complex task indeed. Heat transfer to the body depends on radiation from flames, invisible gases and walls, and on convection currents and thin but nevertheless significant gaseous boundary layers whose thickness is difficult to determine.

Even more difficult is the case where a solid body has a molten core. This type of problem occurs in the metals industry where, after refining, metals are cast into large ingots that will subsequently be hot-rolled into bars. The ingots need to have an approximately uniform temperature throughout, before rolling can commence. When they are removed from the moulds, the ingots are relatively cool on their surfaces with highest temperatures in the molten centres. They are transferred to heating furnaces where their surfaces are heated both externally from the furnace and internally by conduction from the molten cores. The molten core shrinks in size as the process proceeds, and calculation of the temperature distribution needs to take into account the latent heat release as the core solidifies. Fig. 12.6 illustrates the problem schematically.

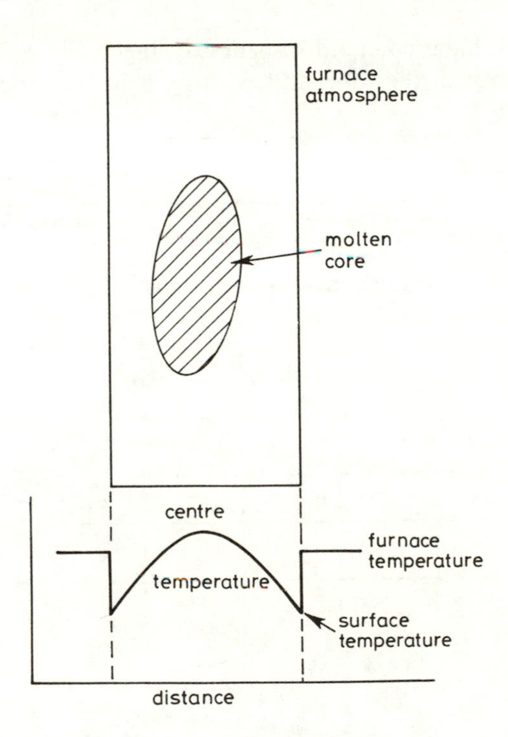

Fig. 12.6 *Heating of a metal block that has a molten core*

Of course, many industries operate repetitively and thus build up a large store of experience and knowledge that can be expressed as a set of 'rules'. Observance of the rules ensures that, within certain limits, heating objectives will always be obtained. Note carefully, though, that many of the rules used by industry are inequalities; e.g. a given minimum heating time is required in order to achieve uniform temperature distribution in a solid of a particular type and size. One energy-conservation measure in such situations is to use improved process monitoring in conjunction with scientific analysis of the heating process to sharpen the inequalities: i.e. to ensure, for the example just quoted, that enough, but no more than enough, heating time is allowed to achieve uniform temperature distribution.

The electrically based methods of heating (induction, direct current, dielectric, microwave) all generate energy internally in solid bodies, and therefore, by their use, more efficient control of temperature distribution may be obtained. The characteristics of each of the above methods in relation to the heating of solids were briefly discussed in Chapter 9.

12.4 Mathematical modelling of a counterflow heat exchanger

Fig. 12.7 shows a tubular counterflow heat exchanger. Fluid in the central tube exchanges heat through the tube wall with fluid in the outer, concentrically

Fig. 12.7 *Tubular concentric counterflow heat exchanger*

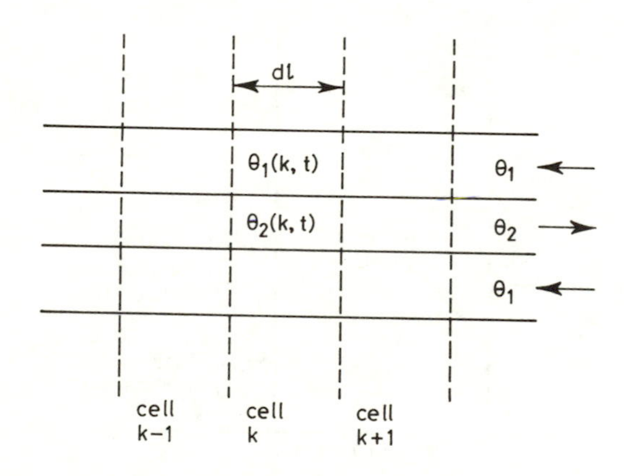

Fig. 12.8 *Modelling starts from an assumption that the heat exchanger may be approximated by a sequence of cells*

surrounding tube. Modelling starts from the assumption (see Fig. 12.8) that the heat exchanger may be approximated by a sequence of cells numbered $i = 1, \ldots, n$, each cell being of length dl. Within each cell, the contents are assumed to be perfectly homogeneous. Let the temperature in the outer tube be θ_1 and that in the inner tube be θ_2; then we may write the outer and inner temperatures as $\theta_1(k, t)$, $\theta_2(k, t)$ for the kth cell.

We assume incompressible fluid flow at mass flow rates W_1, W_2, specific heat of fluids C_1, C_2, densities E_1, E_2, cross-sectional areas A_1, A_2, diameter of inner tube D and mass flow rates to be W_1, W_2, respectively. These quantities are assumed to remain constant. We ignore lengthwise diffusion and assume that temperature in the fluid streams will be governed by heat transfer through the wall separating the two fluids and by lateral flow in both fluids. Considering only heat flow between the two fluids in cell k we have

$$dq(k, t) = \alpha\pi D(\theta_2 - \theta_1)\, dl \tag{12.10}$$

where α is a coefficient governing heat transfer across the boundary between the two fluids. In practice, α represents the effect, not only of heat conduction through the pipe walls, but also of the thin but important layers of near-stationary fluid that lie on both sides of the boundary wall and effectively increase its thermal resistance. The thickness of these near-stationary boundary layers is difficult to determine except from experimental data, but experience has shown that the thicknesses are important and flow-rate dependent.

The volume of the cells are $A_1\, dl$, $A_2\, dl$, respectively, and so we can write, ignoring fluid flow effects for the moment,

$$dq = C_1 \varrho_1 A_1\, dl \frac{d\theta}{dt} = -C_2 \varrho_2 A_2\, dl \frac{d\theta_2}{dt} \tag{12.11}$$

Considering next only the effects of flow. In time dt a mass of fluid $w\, dt$ enters each cell and a similar mass leaves the cell. The temperature of the kth cell is therefore given by (for stream 1)

$$\theta_1(k, t + dt) = \frac{(A_1 \varrho_1\, dl - w_1\, dt)\theta_1(k, t) + w_1\, dt\, \theta_1(k - 1, t)}{A_1 \varrho_1\, dl} \tag{12.12}$$

And if we set

$$\frac{\partial \theta_1}{\partial t} = \frac{\theta_1(k, t + dt) - \theta_1(k, t)}{dt} \tag{12.13}$$

and

$$\frac{\partial \theta_1}{\partial l} = \frac{\theta_1(k, t) - \theta_1(k - 1, t)}{dl} \tag{12.14}$$

Then we obtain

$$\frac{\partial \theta_1}{\partial t} = \frac{w_1}{A_1 \varrho_1} \frac{\partial \theta_1}{\partial l} \tag{12.15}$$

Combining the equation representing heat transfer through the interface between the fluids with the equations for modelling the effect of lateral flow, we obtain, after simplifying,

$$C_1 h_1 \frac{\partial \theta_1}{\partial t} = -w_1 C_1 \frac{\partial \theta_1}{\partial l} - q(\theta_1, \theta_2) \tag{12.16}$$

$$C_2 h_2 \frac{\partial \theta_2}{\partial t} = -w_2 C_2 \frac{\partial \theta_2}{\partial l} + q(\theta_1, \theta_2) \qquad (12.17)$$

where h_1, h_2 is the mass of fluid per unit length and q is the rate of heat transfer per unit length.

If we combine these equations into a single second-order equation it will be found to have the form

$$a_1 \frac{\partial^2 \theta}{\partial t} + a_2 \frac{\partial^2 \theta}{\partial t\, \partial l} + a_3 \frac{\partial^2 \theta}{\partial t^2} + a_4 \frac{\partial \theta}{\partial t} + a_5 \frac{\partial \theta}{\delta l} = 0 \qquad (12.18)$$

and it can be seen to be a hyperbolic partial differential equation. Numerical methods for solution of such equations can be found in specialist texts such as Bellman (1985).

12.5 Classification of partial differential equations

Elliptic partial differential equations have solutions that are determined at every point within a closed contour by the boundary value on that contour (Fig. 12.9).

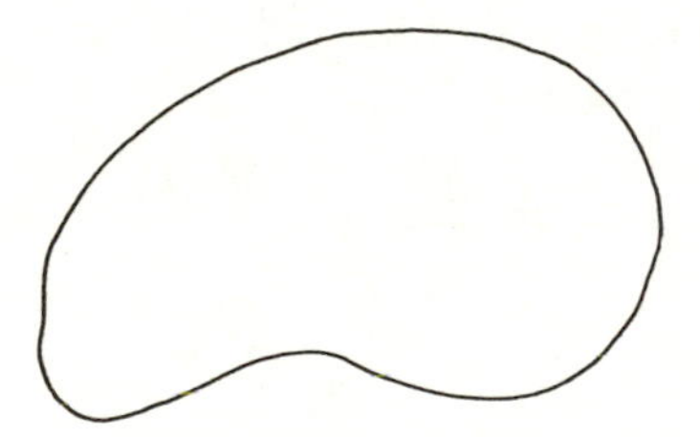

Fig. 12.9 *Elliptic partial differential equations determine the solution at every point within a closed contour*

Initial conditions are given at every point within the contour. The time solution describes how the initial conditions evolve into final conditions that are in equilibrium with the boundary conditions on the contour.

Parabolic equations have solutions in a semi-infinite strip (Fig. 12.10). Initial conditions are given at $t = 0$ and boundary conditions for $t \geqslant 0$. A typical problem involving a parabolic equation is that in which a long metal bar is in contact at each end with a heat reservoir, and it is required to determine the temperature distribution as a function of time.

Hyperbolic partial differential equations again have solutions in a semi-infinite strip, but, compared with parabolic equations, additional phenomena are modelled. The propagation of the solution is governed by lines called *characteristics* that define the region of influence of the boundary conditions (see

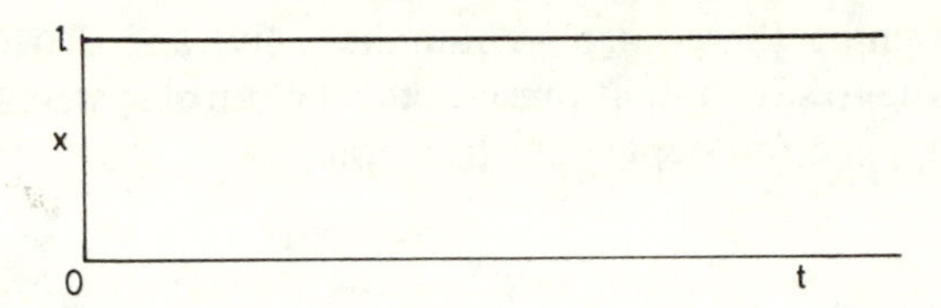

Fig. 12.10 *Parabolic partial differential equations have solutions in a semi-infinite strip*

Figure 12.11). Hyperbolic equations are difficult to solve numerically because of discontinuities at the characteristics. Hence they are often replaced by approximating parabolic equations during practical studies of, for instance, heat exchangers.

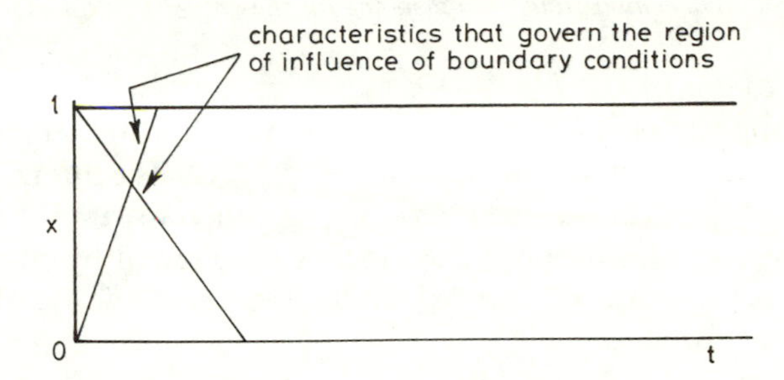

Fig. 12.11 *Hyperbolic equations determine the solution in a semi-infinite strip*
The boundary conditions along the lines $x = 0$ and $x = l$ affect the solution only to the right of the appropriate characteristic

12.6 Control of distributed-parameter problems

A problem where temperature distribution (rather than mean temperature) has to be controlled is called a distributed parameter problem. In a distributed parameter problem, the following sub-problems arise:

(i) What temperature distribution are we aiming for and how should it be required to evolve as a function of time?
(ii) How can the distribution of (i) be described quantitatively; e.g. by some approximating function or by a set of numbers forming an approximating vector?
(iii) How can the actual temperature distribution be measured using available sensors or be estimated mathematically using practicable algorithms?
(iv) How can the temperature distribution be influenced, modified or forced to conform to the specifications decided in (i)?

Topics (i) and (ii) are heavily application dependent and (iii) is partly application dependent. The requirement in (ii) is met by a number of mathematical modelling techniques of differing degrees of rigour and exactness.

Means of attaining the control requirement (iv) are also discussed below. However, it is true to say that, at present, most control systems are designed on fairly obvious application-dependent principles.

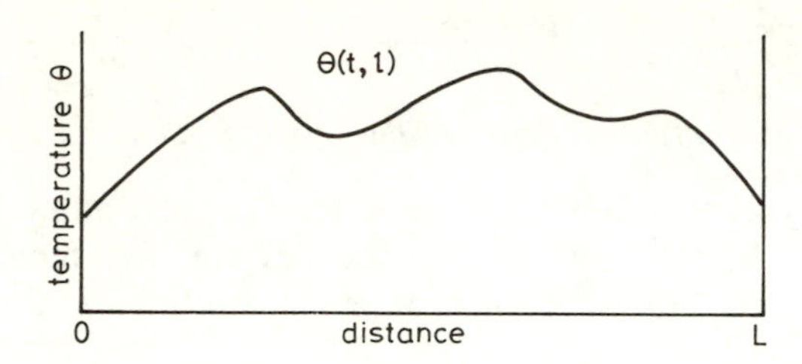

Fig. 12.12 *Variation of temperature with distance for the modal-control example*

12.6.1 Modal control

Consider an application in which the spatial variation of temperature with distance as l varies from 0 to L is as given in Fig. 12.12. The information in Fig. 12.12 will be obtained in practice by the use of either a scanning sensor or by the use of a number of sensors at fixed locations followed by curve fitting.

Modal control consists in approximating the curve of Fig. 12.12 by an expression of the form

$$\theta(t, l) = \sum_{i=0}^{n} \alpha_i(t) f_i(l) \tag{12.19}$$

in which the function f_i are basis functions for the function space to which $\theta(t, l)$ belongs.

The f_i should preferably form an orthogonal set, and sinusoidal functions are obvious condidates. However, a difficulty arises in the application of standard Fourier techniques unless $\theta(t, 0) = \theta(t, l)$ for all t of interest. Other sets of orthogonal functions (Laguerre functions, Legendre polynomials) may be used in the approximation of $\theta(t, l)$ or, alternatively, any set of basis functions may be used for the $\{f_i\}$. However, in the absence of orthogonality, highly inconvenient interaction occurs between terms in the summation of eqn. 12.19.

To illustrate what is meant by interaction between terms suppose that an ordinary power series is used to represent $\theta(t, l)$ with the origin at $\varrho = L/2$, with $L = 1$. The representation is then (at some fixed time)

$$\theta(l) = \alpha_0 + \alpha_1 l + \alpha_2 l^2 + \alpha_3 l^3 + \alpha_4 l^4 + \ldots$$

Consider now the functions $f_2 = l^2$ and $f_4 = l^4$ on the interval $[0, \frac{1}{2}]$.

The 'angle' a between these functions (90° would mean that the functions are orthogonal) is given by

$$a = \cos^{-1}\left\{\frac{|\langle f_2, f_4\rangle|}{\|f_2\|\ \|f_4\|}\right\} = \cos^{-1}\left\{\frac{\left|\int_0^p l^2 l^4\, dl\right|}{\left(\left|\int_0^p l^4\, dl\right| \left|\int_0^p l^8\, dl\right|\right)^{1/2}}\right\}$$

$$= \cos^{-1}\left\{\frac{l^7/7|_0^p}{(l^5/5|_0^p\, l^9/9|_0^p)^{1/2}}\right\} = \cos^{-1}\left\{\frac{p^7/7}{p^7/\sqrt{45}}\right\}$$

$$= \cos^{-1}\left\{\frac{\sqrt{45}}{7}\right\} = 16{\cdot}6° \qquad (12.20)$$

where $\langle\ ,\ \rangle$ indicates inner product and $\|\quad\|$ indicates norm [a generalisation of the length of a vector]

and we see that the two functions f_2, f_4 are strongly interacting to an extent that will lead to ill-conditioned curve fitting (two temperature-distribution curves that differ only by a small amount may have widely differing coefficients).

In considering how many functions are needed to obtain the satisfactory approximation of the temperature distribution (in other words, in considering the choice of n in eqn. 12.19) it is necessary to know the type of control actions that will be taken. Usually there will be little point in modelling those high-frequency local temperature deviations that cannot be modified by control action. Thus, in a typical industrial application, n in eqn. 12.19 may be 4, so that the final closed loop control system will be typically as in Fig. 12.13.

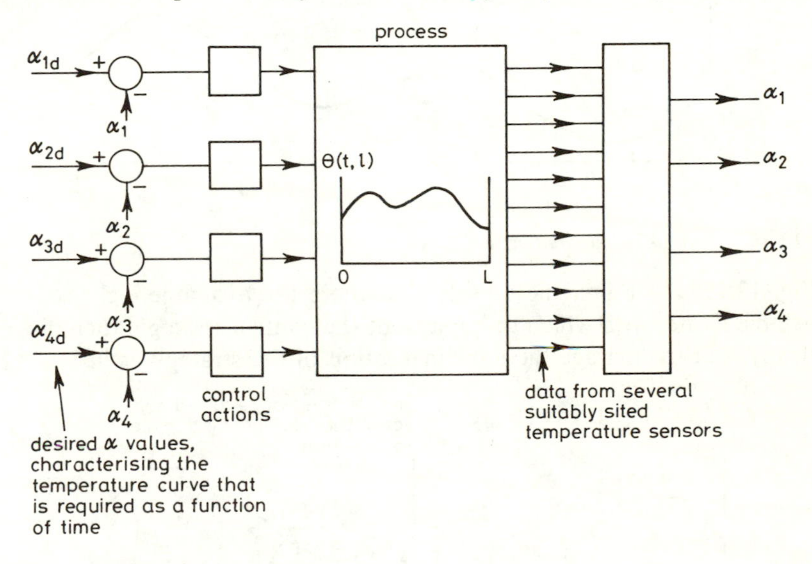

Fig. 12.13 *Modal-control illustration*

The approach as described has been used successfully in controlling flatness of strip from metal rolling mills (Leigh, 1986), but for most temperature-control applications some modification needs to be built in to take account of the lack of truly distributed actuators for temperature control. To illustrate how the implementation may proceed we consider an application where the strategy cells for the control actions $u_0 = \beta_0$, $u_1 = \beta_1 \sin l/\pi$, $u_2 = \beta_2 \sin 2l/\pi$ for l in $-L/2$ to $L/2$ with $\beta_0 = 2$, $\beta_1 = 1$, $\beta_2 = -1$. These control actions are illustrated in Fig. 12.14.

In the absence of three 'distributed actuators' that can be used to implement the three required actions separately, the recommended way forward is to combine the three control actions into one composite required action curve as

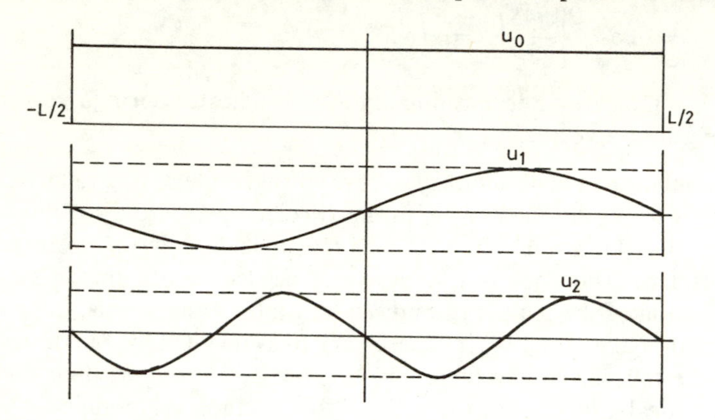

Fig. 12.14 *Set of required control actions*

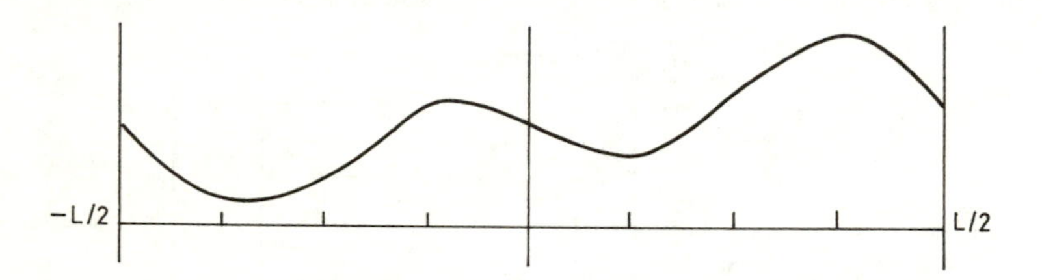

Fig. 12.15 *Composite 'required action' curve*

in Fig. 12.15. If we imagine naively that there are available eight separately adjustable flames with which to implement the control strategy, then diagrammatically we can envisage the implementation of the strategy as in Fig. 12.16.

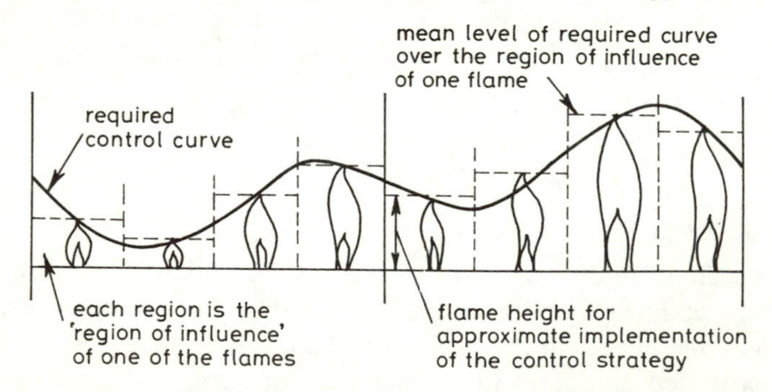

Fig. 12.16 *Implementation of the 'required action' curve of Fig. 12.15*

The required control strategy is averaged over each 'region of influence' (see Fig. 12.15) and each of the eight flames is adjusted according to the mean level of the required control within its region. Of course, this explanation is over-simplified – in particular, the end effects would need special attention; but the approach is clear and is practicable in many real applications.

12.7 References and further reading

BELLMAN, R., and ADOMIAN, A. (1985): 'Partial differential equations: new methods for their treatment and solution' (Reidel Publishing, New York)

JOHN, F. (1975): 'Partial differential equations' (Springer–Verlag, New York)

LEIGH, J. R. (1987): 'Applied control theory' (Peter Peregrinus)

ROOTS, W.K. (1969): 'Fundamentals of temperature control' (Academic Press)

UNBEHAUEN, H., SCHMID, Chr., and BÖTTIGER, F. (1976): 'Comparison and application of DDC algorithms for a heat exchanger', *Automatica*, **12**, pp. 393–402

Chapter 13

Temperature control – 4: Some illustrative examples

13.1 Cascade control for the melting of metals in a small resistance furnace

In a small electrical-resistance furnace for the melting of (initially) solid metals it will often be feasible to use a thermocouple immersed in the molten metal to drive a feedback loop to control the metal temperature. However, if such a system is applied directly, there is a danger that the electrical heating element (shown embedded in the furnace wall in Fig. 13.1) will overheat, particularly in the early stages of a batch when the metal in the furnace is still solid and is therefore feeding back a very low-temperature signal to the controller.

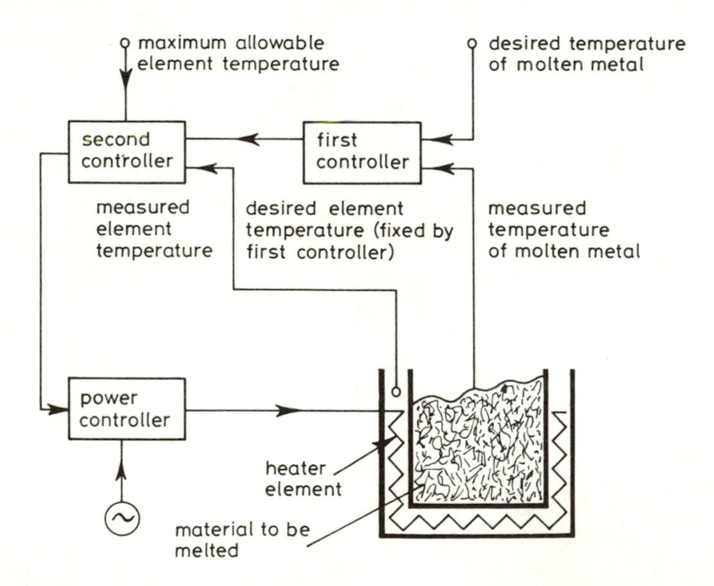

Fig. 13.1 *Cascade control of melting*

A satisfactory solution is shown in Fig. 13.1 where two controllers are connected in cascade and an intermediate signal representing the heating element temperature is connected in. The first controller fixes the desired tem-

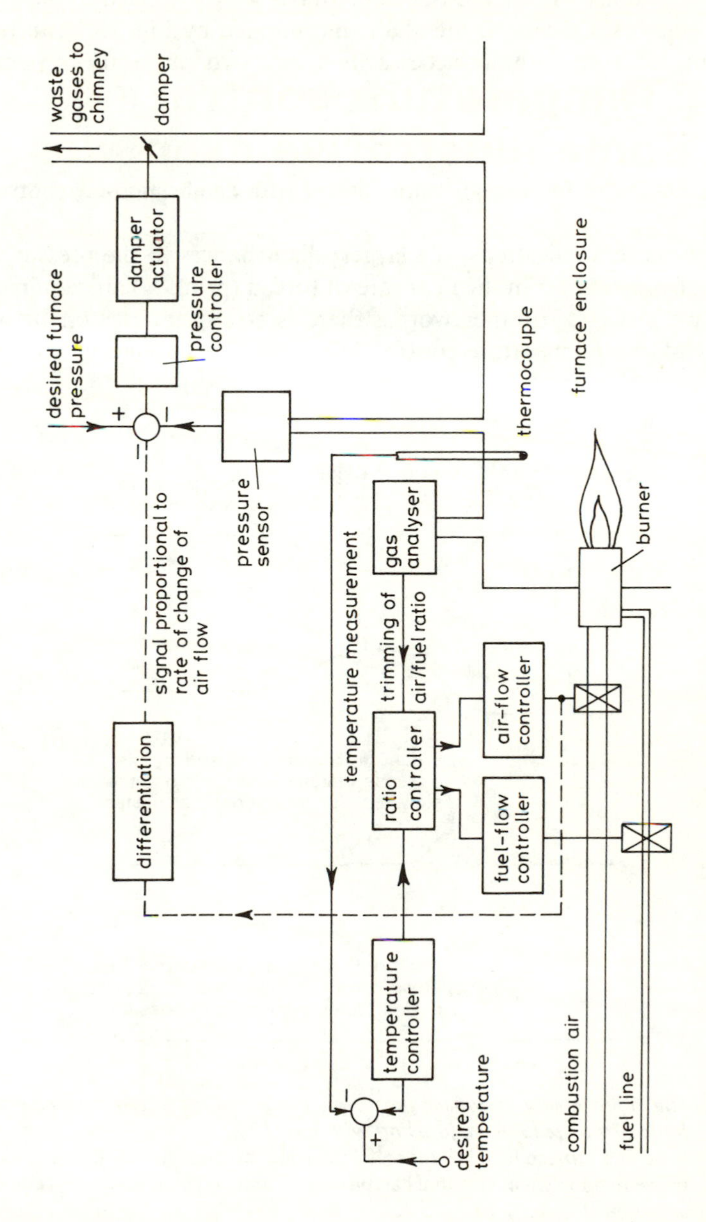

Fig. 13.2 *An example of feedforward control linking combustion air with furnace pressure*
Feedforward loop: When the airflow rate is changing, the pressure controller receives an additional signal

perature for the heating element. The second controller implements this requirement subject to a preset upper limit that is designed to prevent the element overheating. Many standard temperature controllers can be interconnected in the cascade manner shown and the technique is widely applied to the control of jacketed reactors. Of course, the algorithm implied by Fig. 13.1 can be implemented in software without necessarily using two physically separate controllers.

13.2 Interlinking of furnace-pressure control with combustion-air control

In many furnace applications, the largest disturbances to the pressure-control loop arise from changes in the flow rate of forced (primary) air required for the combustion process. In other words, there is strong interaction between the actuators of the temperature-control loop and the measuring sensor of the pressure-control loop.

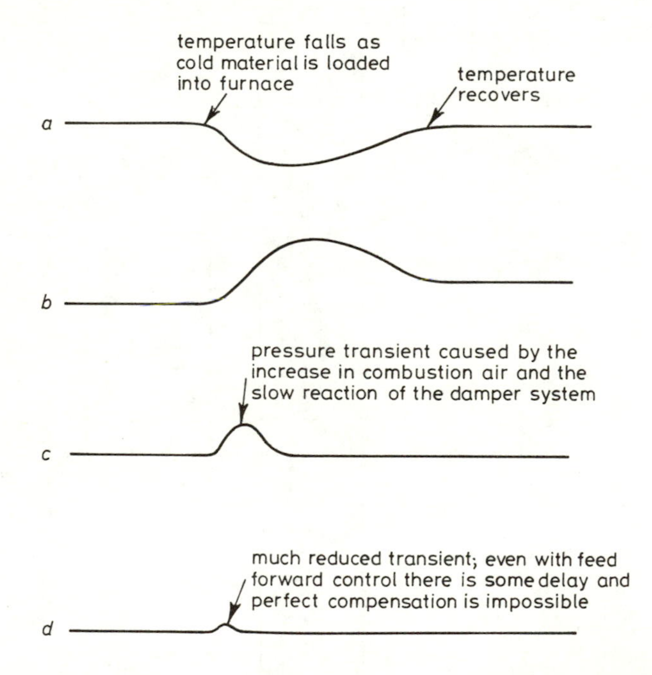

Fig. 13.3 *The improvement in furnace pressure control that feedforward compensation can bring (To be read in conjunction with Fig. 13.2)*
Here, the furnace load is suddenly increased, combustion air flow increases and a pressure transient occurs that can be greatly reduced by feedforward compensation
a Temperature
b Combustion air flow
c Furnace pressure with no feedforward action
d Furnace pressure with feedforward action

Insofar as the interaction is repeatable and calculable, it can be compensated by feedforward action. Basically, instead of waiting for the pressure sensor to signal that a pressure change has occurred and then waiting until the feedback loop has moved a usually heavy damper, we can attempt to link, in some synchronised manner through feedforward action, the movements of the combustion air control with those of the pressure-control damper. In virtually all applications, some trimming feedback control of the pressure-control damper will still be required, since the interlinking through feedforward can never be completely calculable, repeatable or deterministic, and, in any case, furnace pressure is affected by many factors other than the flow rate of combustion air.

Fig. 13.2 shows in outline how feedforward may be added to link combustion air rate with a furnace pressure-control loop while still allowing feedback to correct for inexact feedforward compensation and for other disturbing factors. Fig. 13.3 shows how the pressure control is expected to be improved when feedforward action is added.

13.3 Simple temperature control of a stirred tank (Fig. 13.4)

A heat balance for the system results in the equation

$$P(t) = AH\varrho C\frac{d\theta_1}{dt} + u_2\varrho C(\theta_2 - \theta_1) \qquad (13.1)$$

where P is the rate of heat input from the heater, A, H, ϱ, C are the tank cross-sectional area, liquid height, liquid density and specific heat, respectively

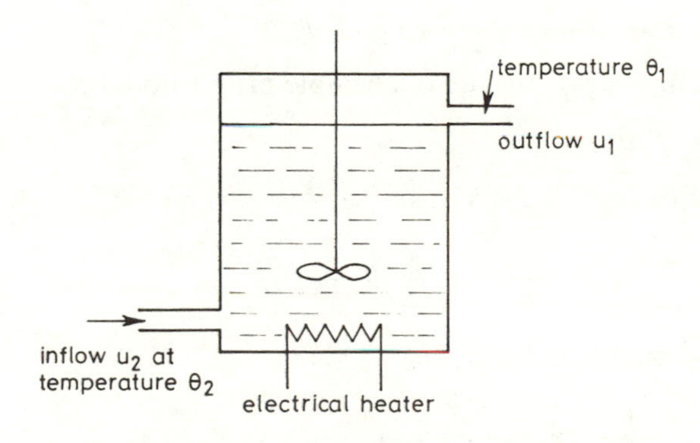

Fig. 13.4 *Stirred tank whose outflow temperature is to be controlled*

– all are assumed constant. Under these conditions, inflow = outflow, i.e. $u_2 = u_2$, and so the right-hand term in the equation represents a net heat loss due to the outflow. The term in $d\theta_1/dt$ completes the heat balance. Consider conditions of constant outflow rate and let $a_1 = AH\varrho C$, $a_2 = u_2\varrho C$; then, equivalent to eqn. 13.1 we can write

Rate of heat gain − Rate of heat loss = Rate of heat input

or

$$a_1 \frac{d\theta_1}{dt} - a_2(\theta_2 - \theta_1) = P(t)$$

$$\frac{d\theta_1}{dt} + \frac{a_2}{a_1}\theta_1 = \frac{a_2}{a_1}\theta_2 + \frac{1}{a_1}P(t) \tag{13.2}$$

We have clearly decided, arbitrarily, that the flow rates u_1, u_2 should be regarded as equal and constant. Now we choose to regard θ_1 as a variable whose value is to be held equal to the desired value θ_d, and to regard changes in θ_2 as disturbances to the system whose effect is to be compensated by feedback control. To achieve feedback control we add a controller D to make $P(t)$ a function of the error $\theta_d - \theta_1$ (see Fig. 13.5). (It is clear that we could equally well have investigated the effect of flow variation or any other aspect of particular interest, but, in this example, we illustrate only one possible development.)

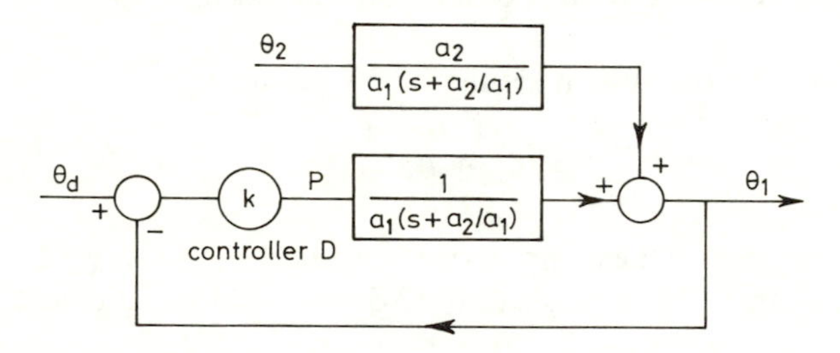

Fig. 13.5 *Stirred tank under closed-loop control*

Initially we investigate the use of a simple proportional controller, i.e. we set

$$P = K(\theta_d - \theta_1) \tag{13.3}$$

The revised equation for the system becomes

$$\frac{d\theta_1}{dt} + \frac{a_2}{a_1}\theta_1 = \frac{a_2}{a_1}\theta_2 + \frac{k}{a_1}(\theta_d - \theta_1) \tag{13.4}$$

Laplace transforming and rearranging yields

$$\left(s + \frac{a_2 + k}{a_1}\right)\theta_1 = \frac{a_2}{a_1}\theta_2 + \frac{k}{a_1}\theta_d \tag{13.5}$$

Fig. 13.5 illustrates the arrangement of the system under closed-loop control. We can see that there are two transfer functions of interest:

G_1, relating θ_1 to θ_d according to the expression

$$G_1(s) = \frac{\theta_1(s)}{\theta_d(s)} = \frac{k/a_1}{(s + (a_2 + k)/a_1)} \tag{13.6}$$

and G_2, relating θ_1 to θ_2 according to the expression

$$G_2(s) = \frac{\theta_1(s)}{\theta_2(s)} = \frac{a_2/a_1}{(s + (a_2 + k)/a_1)} \tag{13.7}$$

Suppose that a step change of magnitude α is made in the temperature θ_2 of the incoming flow, i.e. we have $\Delta\theta_2(s) = \alpha/s$. The response of the system is given by

$$\Delta\theta_1(s) = G_2(s)\Delta\theta_2(s) = \frac{a_2/a_1}{(s + (a_2 + k)/a_1)} \frac{\alpha}{s} \tag{13.8}$$

or

$$\Delta\theta_1(t) = \frac{a_2}{a_1} \frac{a_1}{(a_2 + k)} (1 - e^{-[(a_2+k)/a_1]t})\alpha \tag{13.9}$$

$$= \frac{\alpha a_2}{a_2 + k} (1 - e^{-[(a_2+k)/a_1]t}) \tag{13.10}$$

Similarly for a step change of magnitude β in the desired value θ_d, i.e. $\Delta\theta_d(s) = \beta/s$, the response is given by

$$\Delta\theta_1(s) = G_1(s)\Delta\theta_d(s) = \frac{k/a_1}{[s + (a_2 + k)/a_1]} \frac{\beta}{s} \tag{13.11}$$

or

$$\Delta\theta_1(t) = \frac{k\beta}{a_1} \frac{(a_1}{(a_2 + k)} (1 - e^{-[(a_2+k)/a_1]t})$$

$$= \frac{\beta k}{a_2 + k} (1 - e^{-[(a_2+k)/a_1]t}) \tag{13.12}$$

Notice that the equations we have just derived describe the effect of changes. In the first case, we would wish that $\Delta\theta_1(t)$ was driven quickly to zero to minimise the effect of the disturbance; in the second case, we would wish that $\Delta\theta_1(t)$ quickly arrived at the value $\Delta\theta_d$ to give rapid agreement with the new desired value.

Checking in the equations representing the time solutions we find that:

Case 1 (Step disturbance in incoming temperature): A temperature error of $\alpha[a_2/(a_2 + k)]$ results in the steady state and the condition is approached with a time constant $a_1/(a_2 + k)$. As expected, to obtain a low steady-state error, a high value of gain k is needed.

Case 2 (Step change in desired temperature): In the steady state the change in temperature is $k/(a_2 + k)\Delta\theta_d$ and the condition is approached with time constant $a_1/(a_2 + k)$. Again it is clear that a high gain is desirable to obtain rapid response and low steady-state error.

13.3.1 Addition of an integral term to the controller

We now modify the controller so that it has the transfer function

$$D(s) = k + \frac{1}{sT_I} \tag{13.13}$$

where T_I is the so-called *integral time*. Repeating the analysis with the modified controller we obtain

$$\left(s + \frac{a_2}{a_1}\right)\theta_1 = \frac{a_2}{a_1}\theta_2 + \frac{1}{a_1}\left(k + \frac{1}{sT_I}(\theta_d - \theta_1)\right) \tag{13.14}$$

$$\left(s^2 + \left(\frac{a_2 + k}{a_1}\right)s + \frac{1}{a_1 T_I}\right)\theta_1 = \frac{a_2}{a_1}s\theta_2 + \left(\frac{k}{a_1}s + \frac{1}{a_1 T_I}\right)\theta_d \tag{13.15}$$

and considering only case 2, the effect of a step change $\Delta\theta_2$ in incoming temperature, we obtain

$$\Delta\theta_1 = \frac{\frac{a_2}{a_1}s\Delta\theta_2}{s^2 + \left(\frac{a_2 + k}{a_1}\right)s + \frac{1}{a_1 T_I}}$$

and if $\Delta\theta_2 = \alpha/s$,

$$f = \frac{a_2 + k}{a_1}, \qquad g = \frac{1}{a_1 T_I}$$

$$\Delta\theta_1(s) = \frac{\frac{a_2}{a_1}\alpha}{\left(s + \frac{f}{2} + \sqrt{\frac{f^2}{4} - g}\right)\left(s + \frac{f}{2} - \sqrt{\frac{f^2}{4} - g}\right)}$$

$$\Delta\theta_1(t) = \frac{a_2}{a_1}\frac{\alpha}{-2\sqrt{\frac{f^2}{4} - g}}$$

$$\left\{\exp\left[-\left(\frac{f}{2} + \sqrt{\frac{f^2}{4} - g}\right)t\right] - \exp\left[-\left(\frac{f}{2} - \sqrt{\frac{f^2}{4} - g}\right)t\right]\right\} \tag{13.16}$$

We can see that, as expected, the addition of the integral term in the controller has caused the steady-state value of $\Delta\theta$ to be zero. Again, the higher the value of gain k the faster the response of the system.

Noting that the inverse Laplace transform of

$$\frac{1}{(s + a)^2 + b^2} = \frac{1}{b}e^{-at}\sin bt \tag{13.17}$$

we recognise that, as the integral action in the controller is increased by the use of sufficiently small values of T_I, so the solution becomes increasingly oscillatory. In the tuning of a controller for an actual application, a 3-term controller will be used, with high gain for rapid response, integral action to prevent steady-state error and derivative action to damp oscillations with the actual controller coefficients being carefully chosen to give overall optimum response.

13.3.2 Illustration of the use of feedforward control to counteract the effect of disturbances

The principle of feedforward is to measure incoming disturbances, calculate the expected effect on the system and to take compensatory actions. Feedback control is then left with the (hopefully minor) task of correcting only for the residual errors that are due to imperfect compensation.

Let $\Delta\theta_1$ be the disturbance that is to be compensated. Using knowledge of $G_2(s)$ from eqn. 13.7 with k set equal to zero, we calculate that, if uncompensated by control actions, there will be a corresponding change in temperature θ_1 given by

$$\Delta\theta_1(s) = \frac{a_2}{a_1}\frac{1}{\left(s+\dfrac{a_2}{a_1}\right)}\Delta\theta_2(s) \tag{13.18}$$

Feedforward consists in manipulating the heater power to produce an equal and opposite compensatory change in θ_1. From eqn. 13.2 we can derive the relation

$$\frac{\theta_1(s)}{P(s)} = \frac{1}{a_1\left(s+\dfrac{a_2}{a_1}\right)} \tag{13.19}$$

and hence

$$\Delta\theta_1(s) = \frac{1}{a_1\left(s+\dfrac{a_2}{a_1}\right)}\Delta P(s)$$

For compensation, we require that

$$\Delta P(s)\frac{1}{a_1\left(s+\dfrac{a_2}{a_1}\right)} + \Delta\theta_2(s)\frac{a_2}{a_1}\frac{1}{\left(s+\dfrac{a_2}{a_1}\right)} = 0$$

i.e. we require that

$$\Delta P(s) = -a_2\Delta\theta_2(s) \tag{13.20}$$

If we imagine that the system is set up initially to operate perfectly with an incoming temperature θ_2° and a heater setting P° then feedforward consists of arranging that

$$P(t) = P^{\circ} - a_2[\theta_2(t) - \theta_2^{\circ}] \qquad (13.21)$$

Feedback is still required to reduce to zero the effects of (inevitable) imperfect feedforward compensation (see Fig. 13.6). Notice that feedforward compensation is an open-loop strategy and that there is therefore no danger of dynamic instability because of interaction between 'rival' control loops.

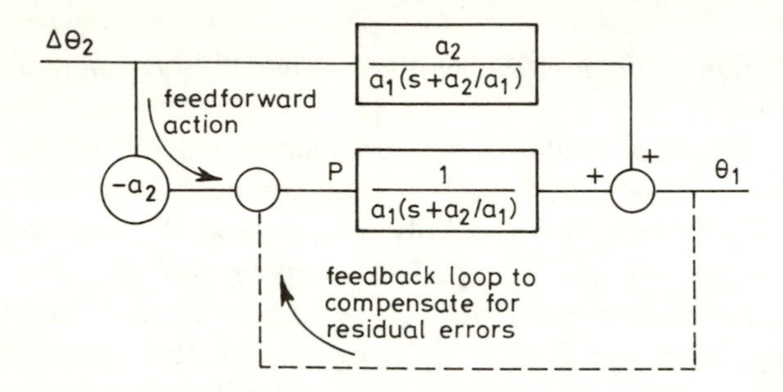

Fig. 13.6 *Illustration of the operation of feedforward action to compensate for changes in the temperature θ_2*

Finally, note that, fortuitously in the example just given, the dynamic effects in eqn. 13.19 cancelled. In general, disturbance effects propagate through one transfer function, say $P_1(s)/Q_1(s)$, and compensation by feedforward action takes place through a second transfer function, say $P_2(s)/Q_2(s)$. A unity disturbance affects the output by an amount $P_1(s)/Q_1(s)$ and so the compensatory action required is $-[P_1(s)Q_2(s)]/[Q_1(s)P_2(s)]$. Such an approach attempts to synchronise the compensation with the effects of the disturbance. Such synchronisation will not always be possible, as, for instance, when the transfer function $[P_1(s)Q_2(s)]/[Q_1(s)P_2(s)]$ is not physically realisable.

In physical terms, this situation will occur whenever an actuator with slow dynamics is required to compensate for a fast-propagating disturbance. In the converse situation (fast dynamics required to compensate for a slowly propagating disturbance), synchronisation consists in deliberately slowing the response of the actuator to match that of the disturbance effect.

Fig. 13.7 *Concept of a regenerator*
The large surface area and high heat capacity extract heat from waste gases. The heat is recovered at a later stage by passing cold air through the regenerator

13.4 Control of regenerators

A *regenerator* is an enclosure, lined with material of high heat capacity and large surface area (Fig. 13.7). For use as a heat-recovery device, high-temperature waste gases are conducted through the regenerator for some optimum period of time. At the end of this period, the waste gases are switched to pass through a second regenerator and heat is recovered from the first regenerator by passing an air stream through it. The heated air obtained in this way is usually used as combustion air for the original process whose waste gases were used to heat the regenerator (Figs. 13.8 and 13.9 show the principles involved). The control problem consists largely in determining the most efficient change-over time when the regenerators should be switched.

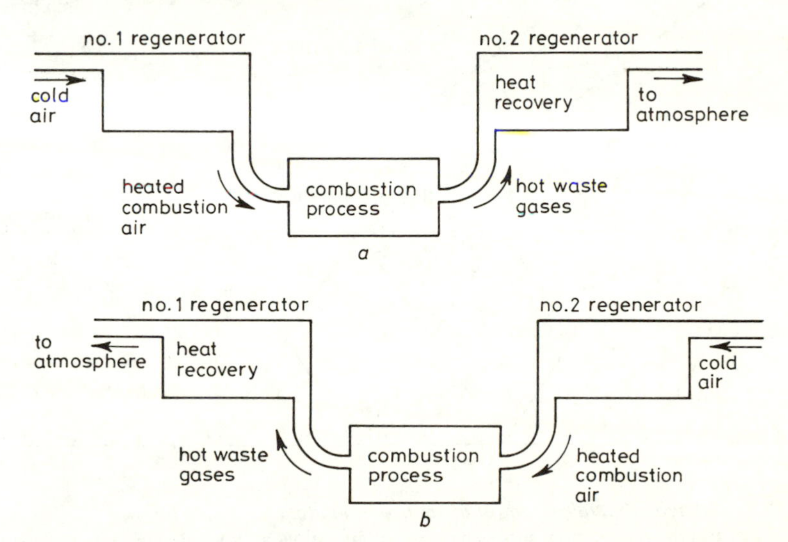

Fig. 13.8a and b *Diagrams illustrating how regenerators are used to preheat combustion air*

A number of metallurgical processes such as smelting processes have a need for large volumes of high-temperature air (> 1000°C) at constant temperature. . These needs are met by using two or more regenerators. One regeneator, A, supplies air at a temperature $\theta_A > \theta_{des}$, where θ_{des} is the desired constant temperature of the hot-air supply. Temperature control of the air supply is achieved by mixing the correct proportions of cold air with the air at temperature θ_A. Let θ_0 be the temperature of the cold air and Q_A, Q_0, Q_{des} be the flow rates of air at temperature A, of cold air and at the desired flow rate, respectively; then control consists in manipulating flow rates to satisfy the equations

$$\frac{Q_A\theta_A + Q_0\theta_0}{Q_A + Q_0} = \theta_{des} \tag{13.22}$$

$$Q_A + Q_0 = Q_{des} \tag{13.23}$$

The other regenerator(s) are being heated while regenerator A is supplying hot air. As soon as $\theta_A = \theta_{des}$, regenerator A must be switched over to the heating phase and another regenerator takes over the task of supplying the hot air (see Fig. 13.8).

The arrangement just described is inefficient because the regenerators have to be heated to temperatures well in excess of θ_{des} in order to allow for dilution by cold air. In return for capital expenditure in additional regenerators, the staggered parallel approach described in Section 13.4.1 offers greater efficiency.

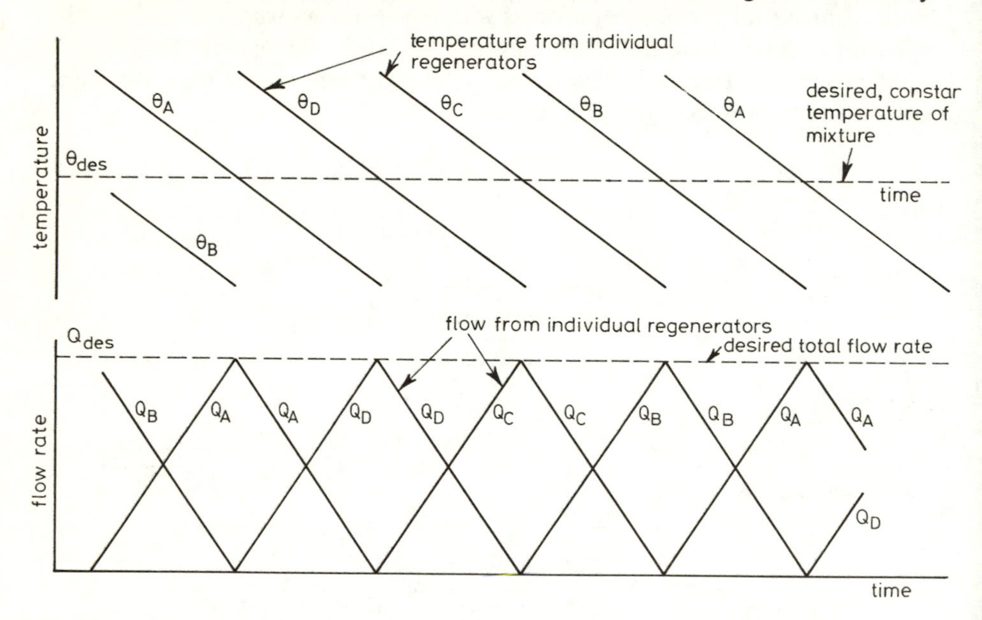

Fig. 13.9 *Staggered parallel control of four regenerators*
Between them they supply a constant air supply at flow rate Q_{des} at temperature θ_{des}. Two regenerators are always supplying heat, with the correct temperature being obtained by continuously altering flow rates from individual regenerators. Two regenerators are always being reheated and the whole arrangement cycles as shown

13.4.1 Staggered parallel operation of regenerators

Suppose we have available four regenerators labelled, A, B, C, D. At any particular time, one regenerator A is supplying air at temperature $\theta_A > \theta_{des}$, one regenerator B is supplying air at temperature $\theta_B < \theta_{des}$ and two regenerators C and D are in their heating phases.

The air-flow rates through regenerators A, B are Q_A, Q_B, respectively, satisfying the relations

$$Q_A + Q_B = Q_{des} \tag{13.24}$$

and

$$\frac{Q_A\theta_A + Q_B\theta_B}{Q_{des}} = \theta_{des} \tag{13.25}$$

Regenerators A, B both cool as they supply air – we assume for the sake of simplicity that this cooling is linear and flow independent. After some time, we reach the situation where $\theta_A = \theta_{des}$. Then the regenerators have their roles changed. D takes the role of air supplier with $\theta_D > \theta_{des}$, A takes the role of air supplier with $\theta_A < \theta_{des}$, B starts its heating phase while C continues its heating phase. Fig. 13.9 illustrates the situation. The arrangement offers improved efficiency compared with the simpler scheme of cold-air dilution because the regenerators do not have to be heated to such a high initial temperature to achieve a particular temperature, θ_{des}.

Chapter 14

Commercially available temperature controllers

14.1 Introduction

When some particular temperature within a process is to be controlled, several choices confront the designer:

(i) Whether to use a commercially available temperature controller.
(ii) Whether to use a more generalised commercially available single-loop process controller.
(iii) Whether to use a multi-loop local controller that can, if necessary, form part of a co-ordinated control system.
(iv) Whether to make up the necessary hardware and software from readily available computing and interfacing modules.

Factors influencing the choice will include:

(i) The presence of other variables that need to be controlled; the need for integration, reliability and for management information.
(ii) The need for standardisation and harmonisation of devices within an area.
(iii) The difficulty of the temperature-control task, the need for non-standard control algorithms, the need for interlinking the temperature-control loop with other control loops.

Broadly the advantages of using a commercially available temperature controller will be ease of application and low cost. The chief disadvantage will usually be the difficulty of integrating and harmonising the temperature-control loop into a larger scheme.

Temperature controllers adequate for many tasks may be purchased for under $100. At these prices, it is possible to obtain plug-in or panel-mounted indicating controllers with on–off or PID algorithms whose parameters are set in through manual switches. A typical basic specification for a low-level temperature controller is:

(*a*) Input designed to receive Chromel/Alumel or other popular base-metal

thermocouple signal with in-built correction for the non-linearity of the specified thermocouple.
(*b*) Cold junction compensation.
(*c*) Thermocouple break-protection (system switches off power on thermocouple becoming open-circuit).
(*d*) On–off or PID algorithm specified by purchaser or sometimes both are provided.
(*e*) High-low low-temperature alarms.
(*f*) Solid-state relay available to drive electrical loads or other on–off actuators.

Additional features that are available in more sophisticated temperature controllers include:
(i) Provision for different controller gains during heating and cooling, to take account of the often widely differing slopes of heating and cooling curves of the process at the operating temperature.
(ii) Provision for programming the temperature set point to follow a particular time trajectory; e.g. to follow a rising ramp, to remain constant for a specified time and then to follow a falling ramp.
(iii) Provision for the control of active cooling as well as of active heating, allowing for either a dead space or an overlap between the two modes.
(iv) Provision for self-tuning, in which the controller chooses its own PID parameters.
(v) Provision for serial data links for connection to a host computer using, for instance, ASC1I protocols to allow for remote supervision and reporting, plus provision for connection of displays, printers, terminals and disc drives.

14.2 Standard low-cost temperature controllers

There are many manufacturers who supply small off-the-shelf controllers specifically designed for control of temperature. Each of these manufacturers offers a range of such devices varying in sophistication and type of display. We list below the outline specification of the MD–SI controller of Calex Instrumentation Ltd. as just one example of a relatively sophisticated off-the-shelf device that can control both heating and cooling.

Controller: MD-J1H6 of Calex Instrumentation Ltd.
Input: Type K or J thermocouple or 100 Ω resistance thermometer
Cold-junction compensation: Built in
Weight: 1·5 kg
Power consumption: 6 W
Heating controller: 3-term PID
Cooling controller: 3-term PID
Overlap or dead space between heating and cooling: Adjustable

Auxiliary relay: Available to switch emergency actions
Outputs: 4–20 mA DC, voltage pulses or relay contact available
Remote operation: Available

14.3 Program controllers

So-called Program controllers are available integrated with temperature controllers. They provide for the automatic programming of the desired temperature setting to follow a desired trajectory that is set up by the use of segments of ramp and constant as shown in Fig. 14.1. When applying a program controller it should be remembered that, even though the desired temperature has been set to follow a given trajectory, it still remains to ensure that the process temperature will follow the desired value. For instance, in a well insulated annealing furnace where the load may have been held at a particular temperature for many hours, it is not too easy to obtain accurate following of the downward sloping ramp.

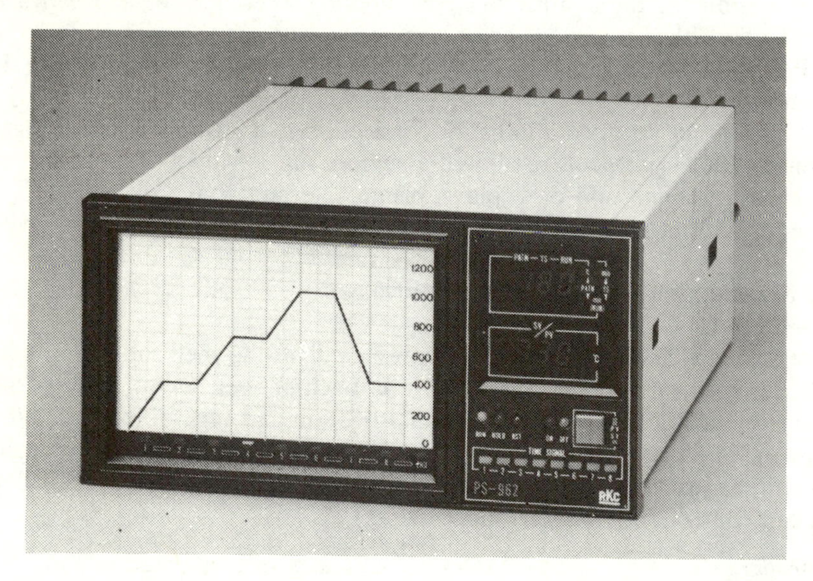

Fig. 14.1 *Program controller (Calex Ltd)*

14.4 Adaptive temperature controllers

It will be an obvious advantage if a temperature controller can reliably calculate and implement optimum settings for itself, both during initial commissioning and periodically to take into account inevitable changes in the process.

The theories of adaptive control and self-tuning regulators have existed for many years and they provide for a range of possible approaches. However, only recently have routinely usable adaptive controllers begun to be available. They include:

(i) *The E5K temperature controller of IMO precision controls:* The E5K controller is a microprocessor-based temperature controller that automatically determines its own PID parameters. The controller is first implemented in a test mode. As the temperature rises towards the desired value the controller monitors the transient behaviour and calculates and displays recommended PID settings. Once this procedure is complete, the control mode is selected and this implements the calculated PID settings.
(ii) *The Turnbull 6355 auto-tuning controller:* The 6355 is a single-loop self-tuning controller intended for a wide class of process variables, including temperature. It operates as shown in Fig. 14.2. Data from the process are fed to an identification element that produces a mathematical model of the process. A design procedure is then followed within the controller to determine optimum settings for the three PID coefficients. These coefficients are displayed, together with a confidence factor. If the confidence factor is sufficiently high, the process operator implements the recommended coefficients by the touch of a button.

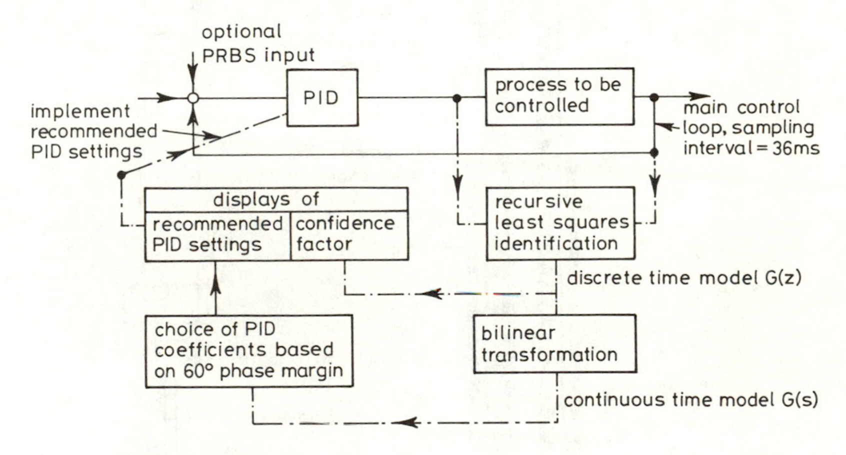

Fig. 14.2 *Turnbull adaptive controller*

14.5 Self-acting temperature controllers

Self-acting temperature controllers rely on the expansion of an oil- or wax-filled system to operate a valve – they need no external power supply. Thus they offer the possibility of simple, reliable, low-cost control systems.

The volume increase of the fluid in the temperature sensor is transmitted through a capillary tube to an actuator where its thrust operates a valve. A

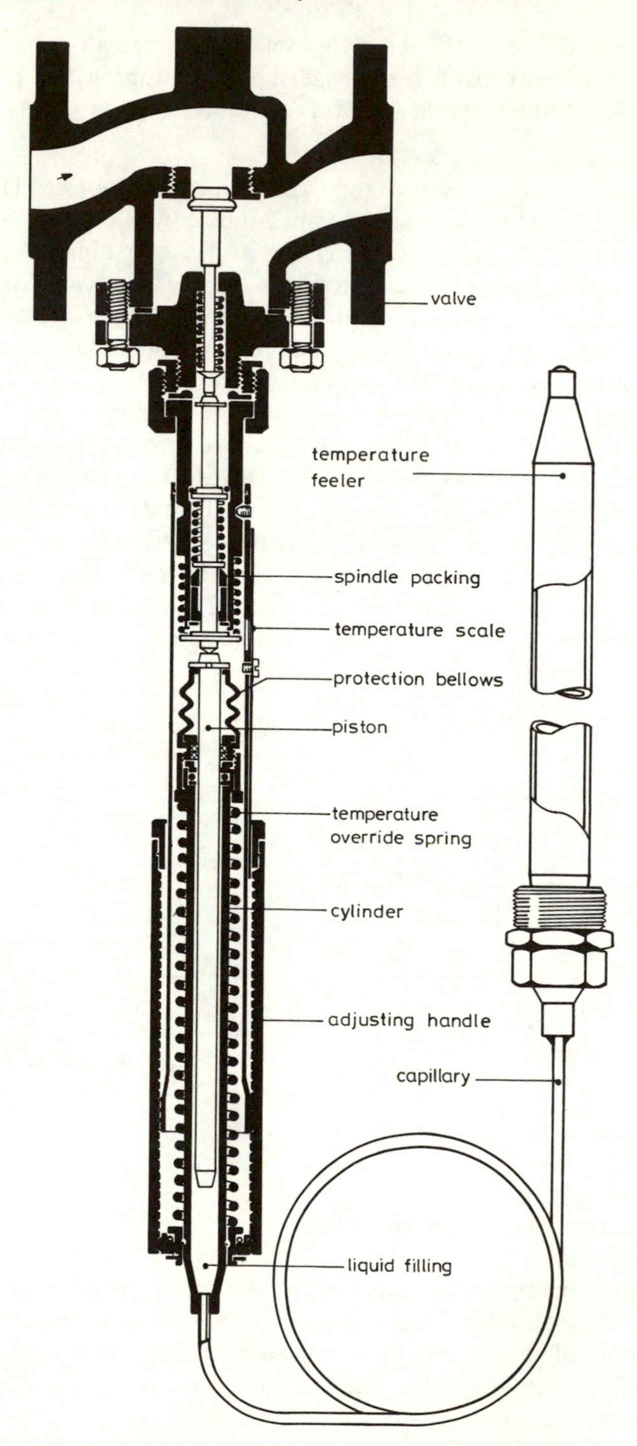

Fig. 14.3 *Self-acting temperature controller (Geestra Ltd.)*

return spring moves the valve in the opposite direction during temperature fall. The behaviour is reasonably linear, providing proportional control which is well suited to many applications.

Controllers may be obtained to control temperatures in the range −15° to 160°C, and the valves manipulate steam or hot water for heating applications or cold water for cooling applications. Fig. 14.3 shows the basic arrangement. The bulbs may be of different forms according to whether the requirement is to control the temperature of a gas or a static or moving liquid. In use, the valves move from open to closed position over a proportioned band of about 10 deg C to control temperatures up to 120°C. For straightforward applications involving the flow of hot water, steam or air, self-acting control valves offer a simple self-contained solution.

Chapter 15

Temperature control in buildings

15.1 Introduction

It is probably true to say that temperature control in most office blocks and similar buildings is unsatisfactory or even very unsatisfactory. Modern office blocks with their large areas of glass are generally worse than older buildings in this respect. To achieve efficient acceptable temperature control in a large building, despite seasonal, diurnal and random changes in external climates, and to provide for a complex occupancy pattern is a very difficult problem. It is one that has remained low on the list of priorities, and relatively little development effort has been invested into improving systems.

A most complex aspect is to quantify and then satisfy human requirements for a comfortable working environment. These requirements cannot be met satisfactorily, as will be seen below, by monitoring and controlling only air temperature. We first consider how the body controls its internal temperature and then proceed to discuss the concept of effective temperature and argue that it is more meaningful than air temperature.

15.2 Temperature control of the human body

The body produces energy from carbohydrates by a process of oxidative metabolism. The energy is needed to drive muscular contraction, glandular secretion and neuronal activity. All of the energy eventually ends as heat energy.

The *rate* of energy conversion depends strongly on the degree of activity of an individual. The lowest rate of energy conversion occurs during sleeping, and rises by a factor of 15 times or more during vigorous physical activity. The heat is produced at a relatively constant rate in the organs of the body and at an activity-dependent rate in the muscles.

Heat loss from a human body obviously depends on the environment, but, as a rough indication, the principal methods by which heat is lost are, in order of decreasing magnitude, radiation, evaporation and convection. Guyton (1974) quotes the respective proportions 60%, 22%, 15% for these quantities under one particular set of experimental conditions. A person may feel uncomfortably cold in a room whose air temperature is high because of cold walls that absorb his radiation.

Evaporative cooling occurs continually under all conditions because of exhalation of water vapour and because of fluid diffusing through the skin. In addition, the sweat glands produce large quantities (up to 4 litres per hour) of fluid when the body is subjected to high environmental temperatures. The evaporation of such large quantities of fluid is the body's chief cooling mechanism, and the only cooling mechanism that remains once the environmental temperature exceeds the body's required temperature. Evaporation rates are highly dependent on the presence or absence of air currents passing over the skin surface. The loss of heat to convective air currents is also highly dependent on the presence of air currents.

15.2.1 Control of body temperature

In the pre-optic area of the hypothalamus is a group of neurons that effectively measure blood temperature. Signals from the neurons pass into the hypothalamus which then responds appropriately by stimulating heat-loss or heat-gain actions in the body.

15.2.2 Heat-gain actions

(i) The blood vessels in the skin are contracted causing less blood to flow in the skin surface, with a consequent reduction in heat loss from the blood to the surroundings. Under this mechanism the skin surface is allowed to cool to a temperature near that of the environment, with consequent reduction in heat loss.
(ii) Hormones are released to increase the metabolic rate.
(iii) The muscles are oscillated to generate heat. In extreme cases this mechanism can be observed as shivering.
(iv) An obsolete mechanism in humans is that the hairs of the body stand on end. This corresponds to a useful mechanism of thickening of fur in animals.
(iv) Thyroid activity increases in the long term (over a period of weeks) the metabolic rate. This mechanism allows a certain amount of useful seasonal adaptation, so that a person becomes acclimatised to lower winter temperatures (provided he is sufficiently exposed to them).

15.2.3 Heat-loss actions

(*a*) The converse of (i) occurs. Blood is directed to the surface of the skin to maximise heat transfer to the surroundings.
(*b*) Sweating is initiated. Fig. 15.1 summarises the heating and cooling mechanisms of the body. Fig. 15.2 shows how heat production decreases and evaporation increases as body temperature rises. The human temperature-control

system has an impressive performance. Exposure for hours on end to temperatures varying over a range of 70 deg C does not displace the central body temperature by more than about 1 deg C.

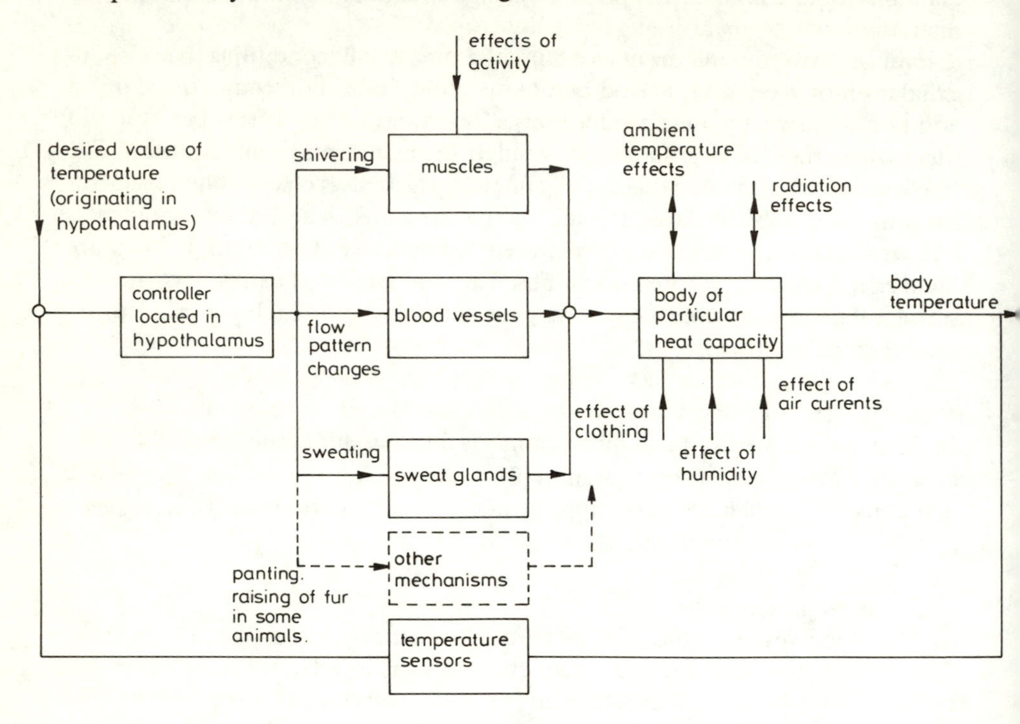

Fig. 15.1 *Simple model of body temperature control*

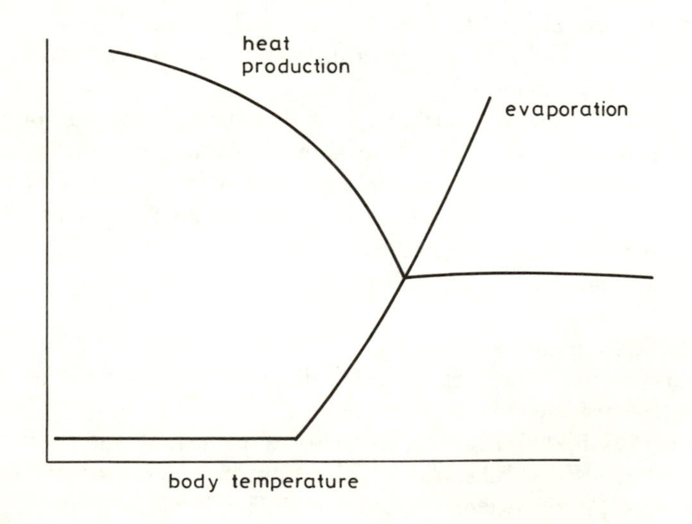

Fig. 15.2 *Heat production and evaporation as a function of body temperature*

15.3 'Apparent' temperature in a room: Concept of effective temperature

It is a matter of common observation that we can feel too hot or too cold in a room whose air temperature is being controlled accurately at (say) 21°C. The reason is that we feel hot or cold dependent on the environmental factors: air temperature, air humidity, air velocity and presence of radiation.

The explanation of these effects is as follows: Humidity affects the rate of evaporation from the skin so that we will feel comfortable at a low temperature if the humidity is high. Air velocity affects both the rate of evaporation from the skin and the convection loss.

To understand the effects of radiation consider returning to a cold house that has been unoccupied for several weeks. The internal air temperature rapidly reaches normal once the heating is turned on, but the house 'feels cold' for some considerable time because we are surrounded by cold walls to which the body radiates strongly. As a further illustration, consider how we feel warm walking on a sunny day in the high mountains despite the low air temperature. Once the sun sinks, we are immediately aware just how much our feeling of warmth was being provided by radiation.

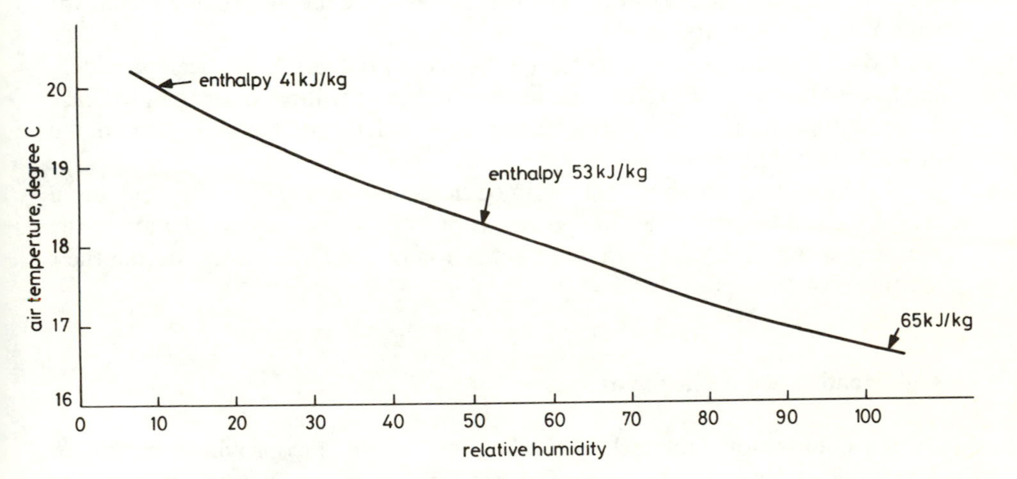

Fig. 15.3 *Combinations of air temperature and relative humidity to yield a constant effective temperature of 17°C*
With indication of enthalpy at some selected combinations

The concept of *effective temperature* attempts to take into account air temperature, humidity and air velocity (but not radiation). The effective temperature of a room at air temperature, relative humidity and mean air velocity $(\theta_A, h, v) = \theta_E$ where θ_E has been defined by multiple subjective experiments as that temperature of dry still air that provides the same temperature 'feel'. In other words we have the subjective equivalence between the combinations

$$(\theta_A, h, v) \quad \text{and} \quad (\theta_E, 0, 0)$$

Once the definition of effective temperature is accepted, the implication is that this quantity, rather than air temperature alone, should be the principal variable to be manipulated in room temperature control.

Fig. 15.3 gives an example of the combinations of air temperature and relative humidity that yield the same effective temperature. It is clear from the Figure that a considerably lower room temperature will be found comfortable by the occupants in a room with high relative humidity. This factor is potentially exploitable to save fuel for heating, but other factors need to be considered.

Consider two cases X and Y in which the same equivalent temperature is provided in different ways. In case X, low-humidity still air is used, and, in case Y, high-humidity still air is used. Approximate values of (θ_A, h, v) for the two cases might be

$$X; \quad (20°\text{C}, 0{\cdot}1, 0)$$

$$Y; \quad (17°\text{C}, 1, 0)$$

We now note three interacting pieces of information:

(i) The enthalpy of a mass of still air at a fixed equivalent temperature is higher for humid air than for dry air. (Thus it requires more fuel to provide the air for case Y than it does for case X).

(ii) However, once the low-temperature humid air for case Y has been provided, the losses through walls to the outside environment will be lower than for case X since such losses depend on temperature differences between θ_A and the outside temperature.

(iii) In the discussion above we assumed zero air velocity, but the conflicting effects (i) and (ii) are in fact weighted according to how rapidly the air in the building is changed; i.e. the rate of ventilation of the building will decide their relative economic importance.

15.4 Heating and cooling systems

During cool seasons, the task of the heating system is to provide a minimum acceptable equivalent temperature in each occupied space in the building at minimum fuel cost.

Heat production by individuals is highly activity-dependent as was seen earlier. Certain work requires individuals to be essentially at rest, while other work involves continuous movement, physical manipulation of objects and frequent walking about. Males have a higher average metabolic rate than females, and tend to wear more clothing and also to be more mobile than their female counterparts. Thus it is difficult to find a single equivalent temperature that will be found acceptable to a group of individuals.

The heating system for a building that is occupied only during mainstream daytime office hours will need to be programmed to switch on at a time in the early morning that depends on the outside temperature and the characteristics of the building plus heating system. At or before the end of the day's occupancy the system needs to be switched off, allowing the temperature to fall to a low level within the safety constraints set by equipment, liability to freezing and residual personnel in the building.

During the hot season, the system needs to cool interior temperatures to the highest acceptable equivalent temperature – minimising fuel expenditure while so doing.

Recalling that heat is generated from human metabolism, lighting, machines and from the effects of sunlight passing through glass, it will often be the case that outside air will be cool enough to be circulated through the building to achieve the required cooling effect. When outside effective air temperature exceeds the upper acceptable limit, it will be necessary to apply refrigeration to the incoming or recirculating air stream.

An ideal control system for a large office building would be highly flexible, allowing individual adjustment by small groups of people engaged in similar tasks. It would have some preprogrammed strategies to allow for occupancy patterns and for seasonal and diurnal climatic changes, with feedback to take account of departures from normality. A number of devices that go some way toward these ideals are becoming available, but, so far, comparatively little effort has been devoted to development in this area. In the longer term, a combination of energy-conscious managements and low-cost control hardware can be expected to make comprehensive temperature control in large buildings commonplace.

15.5 Solar heating of buildings

Solar energy, for heating buildings and for providing a hot water supply, is initially trapped by a flat collector (Fig. 15.4). The energy falls on to a blackened radiation-absorbent surface. Loss of heat from the collector is minimised as shown in the Figure.

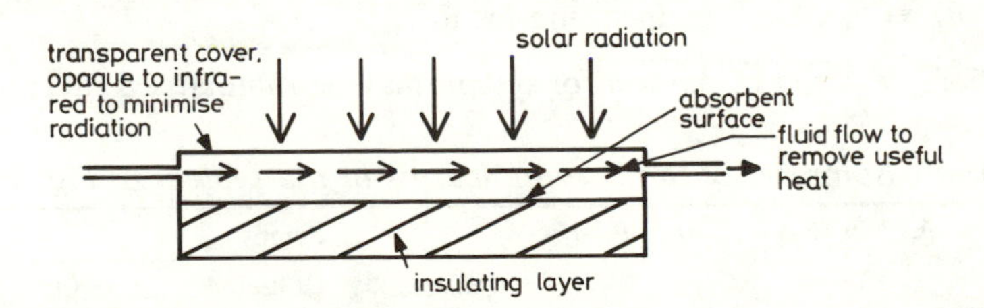

Fig. 15.4 *A solar collector*

In simplified form, a heat pump is applied to the heating of buildings as shown in Fig. 15.5. The system comprises: solar collector, heat store and radiators with ancilliary pumps, sensors and controls. We can consider the state

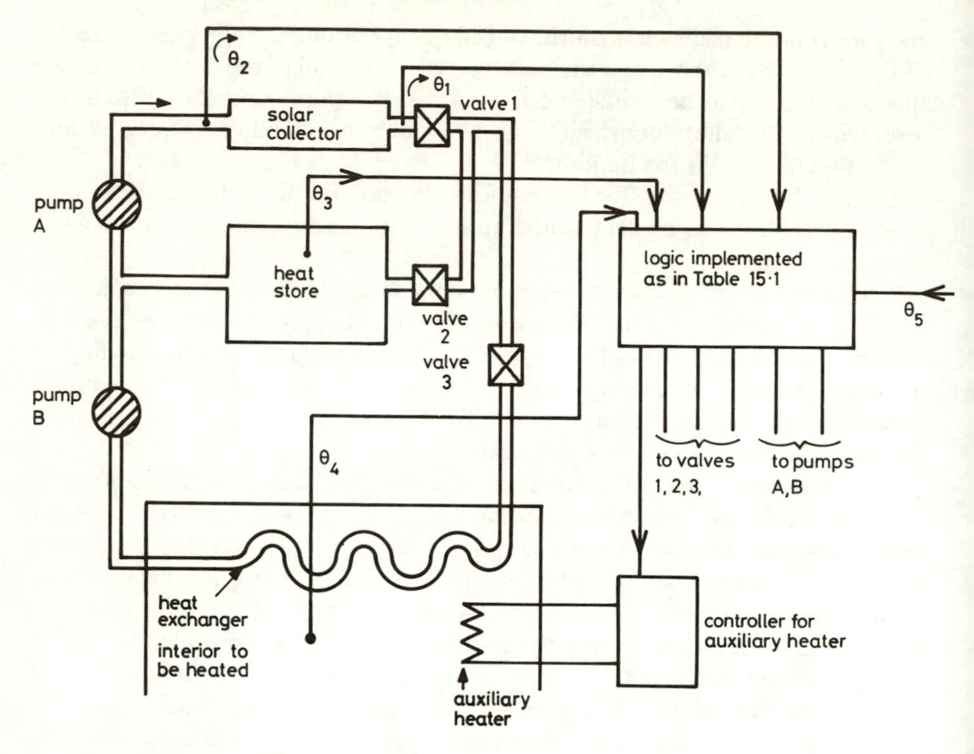

Fig. 15.5 *Schematic showing how a solar collector is used in space heating*

of the system to be characterised by the following temperatures:

θ_1 = temperature of water from solar collector

θ_2 = temperature of water to solar collector

θ_3 = temperature of water in heat store

θ_4 = temperature of environment that is to be controlled

θ_5 = required temperature for θ_4

The actions required of the control system may be summarised in Table 15.1.

Table 15.1 *Summary of the control actions of the system of Fig. 15.5*

$\theta_1-\theta_2$	$\theta_1-\theta_3$	$\theta_1-\theta_4$	$\theta_5-\theta_4$	$\theta_3-\theta_4$	Pumps			Auxiliary heater
					Collector to radiator	Collector to store	Store to radiator	
+		+	+		on			
+	+					on		
		−	+	+			on	
		−	+	−				on

15.6 References and further reading

BENZINGER, T. H. (1982): 'Temperature and thermodynamics of living matter' *in* SCHOOLEY, J. F. (Ed.): 'Temperature: its measurement and control in science and industry', Vol. 5 (American Institute of Physics)

BRUNDRETT, G. W. (1974): 'Controlling the built environment', *Electronics & Power*, April, pp. 248–251

GUYTON, A. C. (1974): 'Function of the human body' (W. B. Saunders)

MAYER, E. (1982): 'Thermal environments and thermal comfort: New instruments and methods' *in* SCHOOLEY, J. F. (Ed.): 'Temperature: its measurement and control in science and industry', Vol. 5 (American Institute of Physics)

Chapter 16

Energy conservation

16.1 Introduction

The initial discussion of the topic of energy conservation in Shinskey (1978) would be difficult to improve upon and the approach below follows his line of reasoning closely.

The energy is an isolated system is constant, but energy can exist in different forms. Although the world is not an isolated system it is clearly incorrect to say that the world's energy is being used up: rather it is being converted into a low-level unusable form. It is the world's fuel resource that is being used up.

Temperature *difference* is the essential driving force of engines or of heating and cooling devices. Hot air is useful for winter heating and cold air is useful for summer cooling, but a mixture of hot and cold air at the same temperature as the environment has no value from a useful energy point of view. As a further example, consider the Atlantic Ocean. It contains a vast quantity of heat energy that cannot be extracted (unless a lower-temperature 'sink' can be found at a deeper level in the ocean or can be provided by some other means such as towing an iceberg southwards). Temperature differences can be exploited to drive useful processes or they can be squandered by irreversible mixing without any useful work being extracted.

Energy-conscious control involves maximum extraction of work from high-level energy sources and/or minimisation of the degradation of energy level along the irreversible route towards inevitable eventual equilibrium with the environment.

The key to application of the approach advocated above is to choose processes and so to control them that entropy increases are minimised. Thus, if two alternative processes are being compared, that with the least associated entropy increase will be the most efficient in preserving energy level. Sometimes a chain of processes will be needed to achieve these goals, as when 'waste' heat from a power station is used to perform useful district heating of local blocks of flats.

A large proportion of our fuel resources are expended on creating order out of disorder in, for instance, manufacture, refining, separation etc. Processes that, overall, minimise entropy increase will be the most fuel-efficient.

The implications of the above include the following:

(i) Unnecessary mixing of materials at different temperatures should be avoided (such mixing is essentially irreversible).
(ii) Unnecessary mixing of previously separated or refined materials should be avoided (again, such mixing is irreversible).
(iii) Temperature differences must be utilised to perform useful work rather than being allowed to reach some intermediate equilibrium without extraction of work.
(iv) Control of flow by constricting an orifice is to be avoided since again this is an irreversible process. Note that there can be no degradation of energy level at a fully closed valve or at a fully open valve. It is in valves in an intermediate position that degradation occurs. The implication is that flow should ideally be controlled by some device such as a variable-speed pump rather than by means of a constricting orifice.

16.2 Irreversible processes

Some of the main irreversible mechanisms that are involved in mechanical, electrical, chemical and thermal processes are:

Stirring of a viscous liquid
Inelastic deformation of a solid
Flow of electric current through a resistor
Magnetic hysteresis
Flow of gas through an orifice or seeping through a porous surface
Chemical reactions
Mixing of dissimilar fluids
Solution of solids in liquids
Transfer of heat from a hot reservoir to a cooler one

16.3 Calculation of entropy changes in typical processes

16.3.1 Calculation of the entropy change due to stirring of liquid in an isolated container

Assuming that the process takes place at constant pressure, let the initial temperature be T_1 and the final temperature be T_2. The entropy change is calculated by postulating an idealised reversible process to take the system from temperature T_1 to temperature T_2:

$$\Delta S = \int_{T_1}^{T_2} \frac{dQ}{T} \qquad \text{where} \qquad dQ = C_p dT$$

$$\Delta S = \int_{T_1}^{T_2} C_p \frac{dT}{T}$$

and if C_p is assumed constant over the range (T_1, T_2), then

$$\Delta S = C_p \ln\left(\frac{T_2}{T_1}\right) \tag{16.1}$$

16.3.2 Calculation of the entropy increase at heat transfer

Let a quantity of heat dq be transferred from a body at temperature T_1 to a body at temperature T_2. The entropy change in the conceptual system shown in Fig. 16.1 is given by

$$dS = -\frac{dq}{T_1} + \frac{dq}{T_2} = \left(\frac{T_1 - T_2}{T_1 T_2}\right) dq \tag{16.2}$$

and clearly $dS > 0$ for any real heat transfer process.

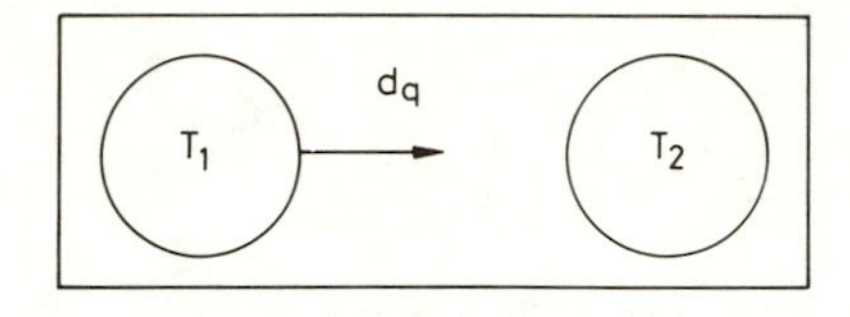

Fig. 16.1 *Heat transfer from a body at temperature T_1 to a body at temperature T_2*

We can imagine that, although no energy has left the system, the capacity for performing work has been reduced and it is this reduction that is quantified by the increase in entropy. Relating this concept to an industrial example, consider a steam superheater (Fig. 16.2) where hot gases from a combustion process are

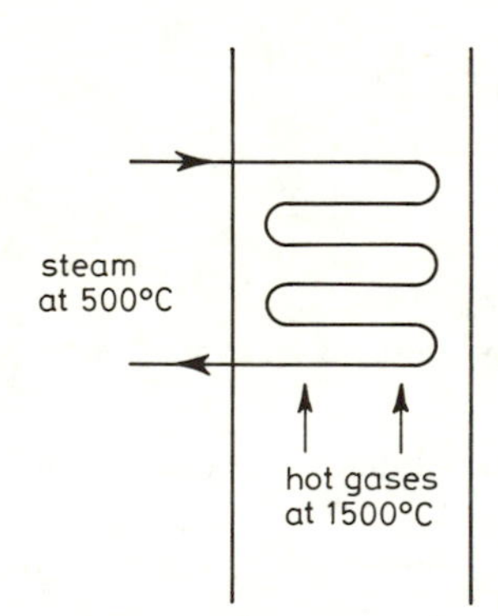

Fig. 16.2 *A steam superheater*

used to heat steam. Insert feasible values of 1500°C for the temperature of the hot gases, 500°C for that of the steam and assume an ambient temperature of 20°C. The theoretical reduction in ability to perform work can be quantified as follows.

An ideal reversible engine operating between upper temperature T and lower temperature T_a is able to convert a proportion $(T - T_a)/T$ (see Section 1.3) of its incoming energy into useful work. The remainder of the energy is rejected at the lower temperature T_a.

Using the temperature values in the steam-raising example we find that, in the hot gases of combustion, $(1773 - 293)/1773 = 0{\cdot}83$, i.e. 83%, of the energy might theoretically have been converted into work, whereas, after the heat transfer, only $(773 - 293)/773 = 0{\cdot}62$, i.e. 62% of the energy is theoretically available to be converted into work.

16.4 Maximum work available from a given quantity of heat energy

Consider a quantity of heat energy q_1 at a high temperature T_1 that passes through some sort of engine (Fig. 16.3) that discharges energy q_2 to a heat sink

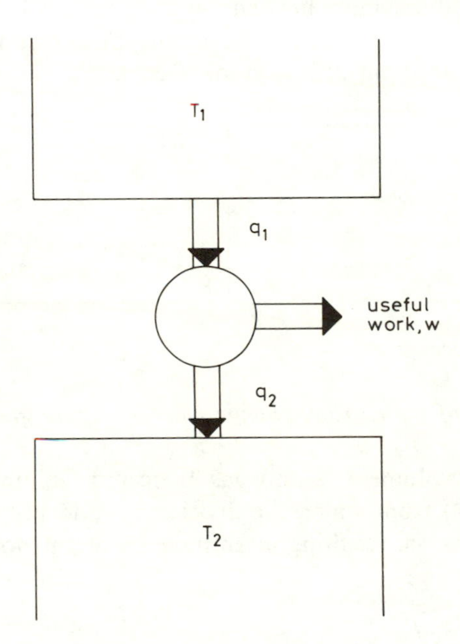

Fig. 16.3 *An idealised heat engine*

at temperature T_2. Maximum work will be made available when there is no overall increase in entropy in the system; i.e. we must have

$$\Delta S_1 + \Delta S_2 = 0$$

or

$$\frac{q_2}{T_2} = \frac{q_1}{T_1} \tag{16.3}$$

From the First Law, energy must be conserved, i.e.

$$q_1 = q_2 + w$$

or

$$\frac{w}{q_1} = \frac{q_1 - q_2}{q_1} = 1 - \frac{q_2}{q_1} = 1 - \frac{T_2}{T_1} = \frac{T_1 - T_2}{T_1} \tag{16.4}$$

and this is the maximum theoretical conversion obtainable from an ideal (non-realisable) engine.

16.5 Conversion of work into heat energy

Whereas heat energy cannot be converted completely into work, work can be converted completely into heat.

A heat pump produces more heat energy than the work that is put in. In this sense a heat pump is more than 100% efficient, but, of course this is explained by the presence of an input of low-grade 'free' energy into the heat pump.

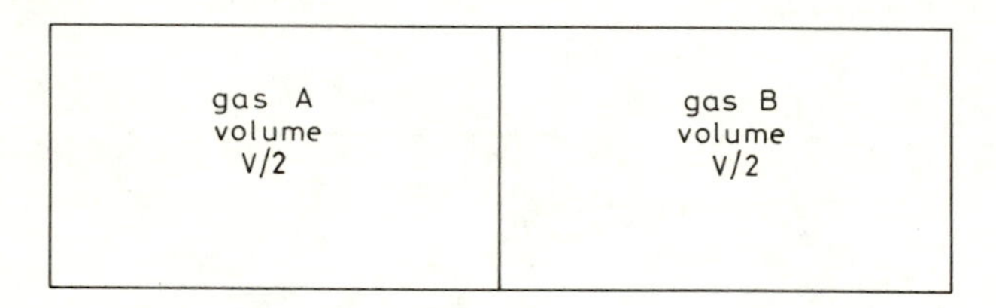

Fig. 16.4 *A divided container*

16.5.1 Calculation of the entropy change when two pure gases are allowed to mix

Let gas A occupy volume $V/2$ and gas B occupy volume $V/2$ in a divided container (Fig. 16.4) from which the division is suddenly removed. Each gas undergoes an expansion, resulting in an increase of entropy for gas A of

$$\Delta S_A = \int_{V/2}^{V} \frac{dQ}{T} \tag{16.5}$$

Assuming an isothermal process,

$$dQ = P\,dV \tag{16.6}$$

For one mole of gas

$$\frac{dQ}{T} = R\frac{dV}{V} \tag{16.7}$$

Hence

$$\Delta S_A = R \ln \frac{V}{V/2} \quad \text{(per mole of gas)} \tag{16.8}$$

and considering the entropy increase due to both gases expanding, we have

$$\Delta S = \Delta S_A + \Delta S_B \tag{16.9}$$

$$= 2R \ln 2 \tag{16.10}$$

We can imagine that separating gases costs work. Allowing pure gases to mix without any extraction of work degrades the energy level of the system as shown by the increase of entropy calculated above. An energy-conscious operating strategy will avoid all such unnecessary mixing of pure materials.

16.5.2 Calculation of the entropy change when a hot fliud is mixed with a cool fluid

Let equal quantities of a fluid at a high temperature T_1 and at a low temperature T_2 be mixed to give a resultant temperature $T_3 = (T_1 + T_2)/2$. The overall change in entropy will be given by

$$\Delta S = \frac{dQ}{T_2} - \frac{dQ}{T_1} \tag{16.11}$$

where Q is the quantity of energy lost or gained by the quantities.

Assuming that the process takes place at a constant pressure and that C_p is constant over the temperature range (T_2, T_1) than, on a per mole basis,

$$dQ = C_p dT \tag{16.12}$$

where

$$dT = T_1 - T_3 = T_3 - T_2 \tag{16.13}$$

Then

$$\Delta S = C_p\left(\frac{dT}{T_2} - \frac{dT}{T_1}\right)$$

$$= C_p dT\left(\frac{T_1 - T_2}{T_1 T_2}\right) = 1{\cdot}5C_p\left(\frac{T_1 - T_2}{T_1 T_2}\right) \tag{16.14}$$

The entropy of the mixed streams exceeds the sum of the entropies of the separate streams by an amount given in eqn. 16.14. This increase in entropy can be considered in terms of an additional fuel requirement to obtain the fluid at T_3 compared with the minimum fuel requirement if temperature T_3 has been obtained by direct heating. The entropy change due to other irreversible processes may be calculated by similar methods (Zemansky, 1981).

16.6 Implications for industrial control of fluid temperature

The analysis in terms of entropy shows that the arrangement of Fig. 16.4 is inefficient compared with the arrangement where the by-pass is removed and temperature control is somehow exercised in the combustion chamber (Fig. 16.5). To understand the physical basis for the inefficiency caused by the bypass it is sufficient to realise that, when the bypass is in use, the flow of fluid through the heat exchanger is reduced and the temperature T_1 rises. Heat transfer from the hot gases to the fluid stream diminishes and the heat hoss in the waste gases rises.

Of course, temperature control using the bypass may be simpler and more accurate than control through the combustion process. This is one of many control design challenges that are generated by the need to conserve fuel.

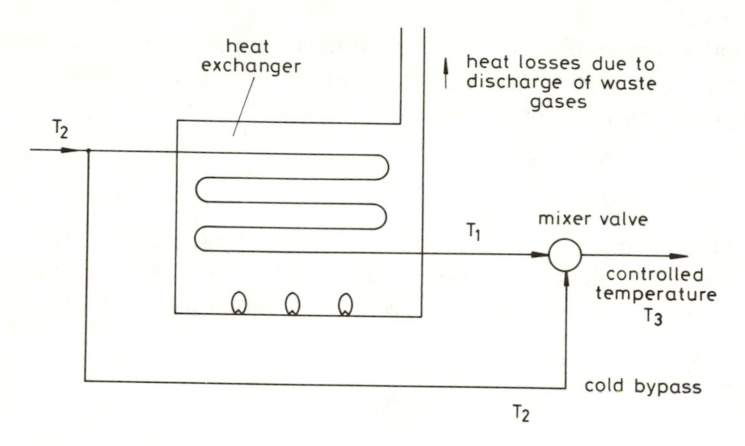

Fig. 16.5 *An arrangement for controlling temperature T_3 by mixing*

16.7 Boiler management systems

Steam-raising boilers are used very widely across industry, and their efficient management is an important ingredient of any factory-wide fuel policy.

A boiler needs to be well instrumented with well-designed control systems enabling it to meet its steam-raising requirement before it can be further developed to achieve a high fuel efficiency. Clearly, tight control, accurate measurement and reporting and good housekeeping will all be necessary requirements in a fuel-efficient operation. In addition, there are a number of specific measures that may be taken to improve boiler efficiency: two of these are described below.

(i) *Minimising energy dissipation across the feedwater control valve:* The flow of cold feedwater to a boiler is conventionally controlled by a valve. Pressure losses across the valve have to be balanced by energy input to the feedwater pump. If a variable-speed feedwater pump is used to control flow, the valve may be removed and the energy loss avoided. This is not quite the end of the story, for the inertia of the variable-speed pump may be too large to allow adequate control to be achieved. In such an eventuality, the valve is retained for the purpose of taking rapid control actions, but the mean level of the feedwater is controlled by the variable-speed pump in such a way that only a necessary minimum energy loss is incurred across the valve.

(ii) *Improved blowdown control:* If water is continually evaporated from a boiler, the concentration of dissolved solids, originally borne in by the feedwater, will increase indefinitely until a stage is reached where deposition or corrosion occurs.

The feedwater to a boiler is treated to reduce the concentration of dissolved solids. Even so, as evaporation of feedwater to steam progresses, the concentration of dissolved solids in the boiler increases steadily. It would, if unchecked, eventually reach a level where deposition or corrosion could occur.

Conventional boiler control therefore incorporates a so-called blowdown strategy: periodically, as dictated by a stored schedule, a quantity of hot water from the boiler is blown to waste. Improved blowdown control involves continuous measurement and inference of the dissolved solids in the boiler contents using conductivity probes, pH probes or composition analysers, with an automatically controlled blowdown operation taking place only as necessary to keep the boiler contents within specification.

Accurate boiler control incorporating features such as those described above will usually rapidly repay the cost of the improved control system and go on to make significant savings in steam-raising costs.

16.8 Implications: Practical guidelines for energy conservation

(i) *Avoid fixing any operating temperature higher than is strictly necessary:* This obvious advice is often blatently disregarded for the reason that operating at a somewhat too high temperature usually makes life easier and avoids the necessity for thinking. For instance, in the heating of metals prior to forming, considerable care is required to calculate and implement the optimum temperature. Too low a temperature may mean that the process may fail or the material crack. Too high a temperature will reduce fuel efficiency and often degrade the product, but it is the easiest strategy for a lazy operator: even a few degrees excess temperature will have a significant adverse effect on fuel efficiency, with high-temperature processes ($>1000°C$) being most sensitive because of the manner in which heat losses increase nonlinearly with temperature. If a liquid is required to be at 60°C, it should be heated to that temperature, rather than being heated to a higher temperature and then brought to the required temperature by mixing with cold liquid.

(ii) *Match heating processes to other interacting processes:* Consider a large furnace that heats metal slabs for input to a rolling mill. The furnace may have a time constant of 20 min and the rolling mill has frequent scheduled and unscheduled breaks in production. Under these conditions it is tempting to let the furnace maintain a constant temperature, so that it can always satisfy its 'customer', the rolling mill. However, simulation will indicate, and implementation will confirm, that a well designed matching of the heating strategy to the mill production schedule will produce very large benefits in improved fuel economy without any associated reduction in production capacity.

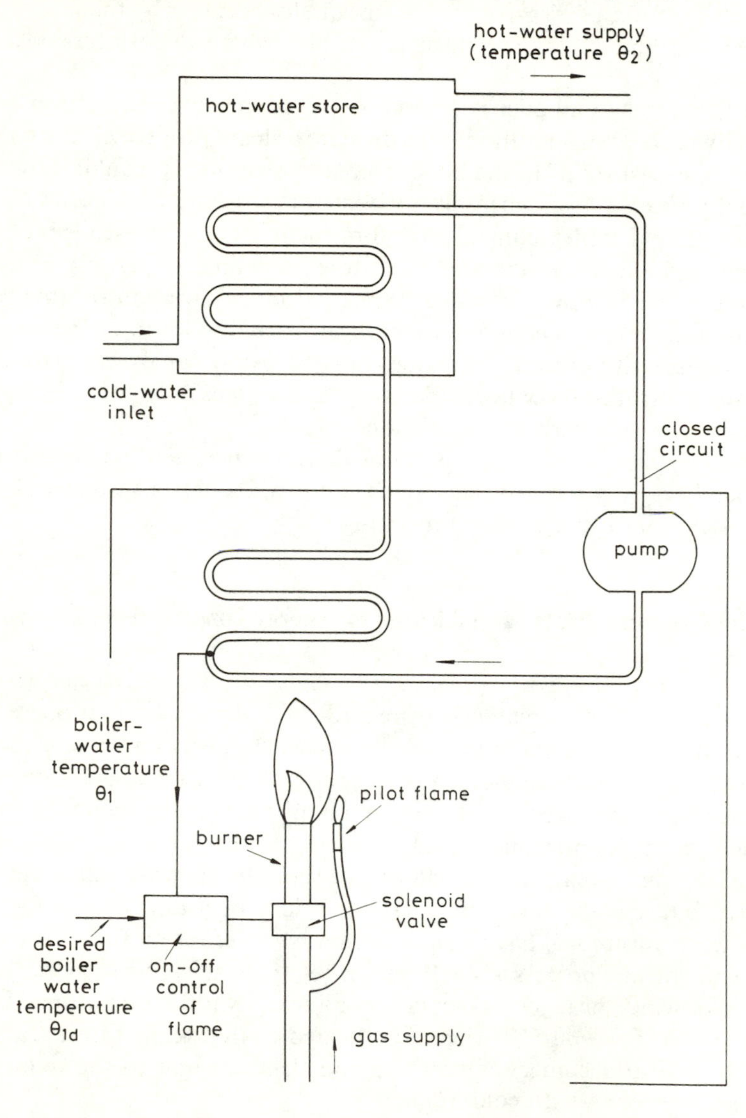

Fig. 16.6 *Simple water-heating system*

(iii) *Ensure correct combustion,* particularly under conditions of light process loading. This point has been adequately covered in Section 8.12).
(iv) *Wherever possible, use waste heat to preheat combustion air.*
(v) *Maintain accurate pressure control,* to prevent ingress of cold air into heated furnaces, buildings or enclosures.
(vi) *Compare carefully the relative merits of alternative strategies.*

(*a*) Consider the simple water-heating arrangement of Fig. 16.6. The aim is simply to provide a constant hot-water supply at a temperature of about $\theta_2 = 50°C$ to be available from the bulk storage tank. The single requirement in this problem is to choose θ_{1d}, the temperature to which the boiler will heat the water that will be circulated by the pump. By choice of a low value for θ_{1d}, we can choose to activate the closed circuit at a low temperature for long periods with short inactive periods between. Alternatively, a high temperature choice for θ_{1d} will result in the closed circuit being activated at a high temperature for short periods with long inactive periods between.

Plotting heat losses against time for the two cases, we shall obtain a graph such as that of Fig. 16.7. A quantitative investigation can be expected to show that there is some intermediate temperature at which losses will be minimised.

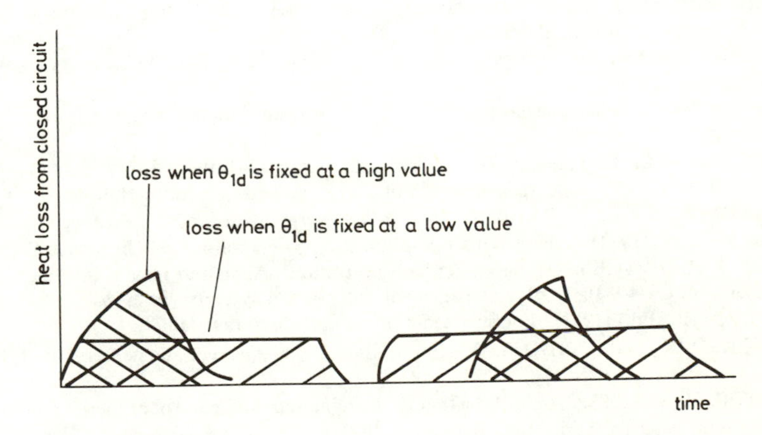

Fig. 16.7 *Comparitive heat-loss profiles for the example of Fig. 16.6 for two different choices of* θ_{1d}

(*b*) A more complex problem: In an (idealised) ironmaking process a high temperature is required, and it can be reached by an appropriate combination of the mechanisms:

(i) The burning of solid or liquid fuel, optionally accompanied by the injection of oxygen gas.
(ii) The deliberate exothermic oxidation of iron (resulting in loss of useful product)
(iii) The injection of steam, which, after dissociation at the process temperature, behaves like a fuel plus oxidant.

Adopting the correct strategy in accordance with changing loading and changing relative costs will result in optimal operation and minimal fuel costs.

16.9 References and further reading

CHO, C. H. (1984): 'Computer-based energy management systems' (Academic Press)

Commission of European Communities (Ed.) (1974): 'Energy conservation in industry. 3 Vols. (VDI)

Conference proceedings

'Energy management in industry'. 12th World energy congress, UNIDO–AIEI, New Delhi, Jan. 1982

'Energy conservation in industry'. Anglo-Swedish Experience Conference, British Institute of Energy Economics, London, Mar. 1982

'Energy conservation in industry'. Meeting Commission of the European Communities, Brussels (3 Vols.) October 1982

'Nonlinear problems in energy engineering'. Symposium, Argonne National Laboratory, Argonne, Ill. Apr. 1983

'Energy management'. 7th National Conference Department of Energy and Industrial & Trade Fairs Ltd., Birmingham, Nov. 1983

8th Annual Conference Energy Management & Controls Society, Cincinnati, Nov. 1982

'Energy efficiency in buildings and industry'. 2nd Mid-Atlantic Energy Conference, Baltimore Gas & Electric Company, Baltimore, Dec. 1983

'Heating equipment and energy conservation'. Symposium, Forging Industry Association, Chicago, Jan. 1984

'Energy economics and management in industry'. European Congress, Algarve, Portugal (3 Vols.) Apr. 1984

DRYDEN, I. G. C. (Ed.) (1982): 'The efficient use of energy' (Butterworth Scientific)

HODGE, B. K. (1985): 'Analysis and design of energy systems' (Prentice-Hall)

KENNEDY, W. J., and TURNER, W. C. (1984): 'Energy management' (Prentice-Hall)

KENNEY, W. F. (1984): 'Energy conservation in the process industries' (Academic Press)

KRENZ, J. H. (1984): 'Energy conversion and utilisation' (Allyn and Bacon, Boston, USA)

MACEDO, M. C. (1983): 'Energy management and control systems' (John Wiley)

PARKER, S. P. (Ed.) (1981): 'Encyclopaedia of energy' (McGraw–Hill)

PATRICK, D. R., and FARDO, S. W. (1982): 'Energy management and conservation' (Prentice-Hall)

PAYNTER, H. M. (1985): 'An introduction to the dynamics and control of thermofluid processes and systems', *J. Dyn. Syst. Meas. & Control,* **107,** pp. 230–232

SHINSKEY, F. G.: (1978): 'Energy conservation through control' (Academic Press)

ZHANG, Q., and SHOURESHI, R. (1985): Multivariable adaptive control of thermal mixing processes', *J. Dyn. Syst. Meas. & Control,* **1-7,** pp. 284–289

General references

BAKER, H., RYDER, E. A., and BAKER, N. H. (undated): 'Temperature measurement in engineering', Vol. 1 and 2 (Omega Engineering, Stamford, USA)

BENEDICT, R. P. (1977): 'Fundamentals of temperature, pressure and flow measurements (John Wiley)

BILLING, B. F., and QUINN, T. J. (1975): 'Temperature measurement' (Institute of Physics) Series No. 26

SCHOOLEY, J. F. (Ed.) (1982): 'Temperature measurement and control in science and industry', Vol. 5 (American Institute of Physics)

EBERHARDT, P., ERLBACHER, J., ERNST, D., and THOMA, M. (1980): 'The control behaviour of temperature controllers based on microprocessors', *in* 'Measurement and automation techniques'. Proceedings of the Interkama Congress, Düsseldorf and Berlin (Springer–Verlag)

FILMORE, R. L. (1982): 'Modelling a closed loop system' *in* SCHOOLEY, J. F. (Ed.): 'Temperature: its measurement and control in science and industry'. Vol. 5 (American Institute of Physics)

LEIGH, J. R. (1987): 'Applied control theory' (Peter Peregrinus)

LEIGH, J. R. (1984): 'Applied digital control: theory, design and implementation' (Prentice–Hall International)

MYLROI, E. G., and CALVERT, G. (1984): 'Measurement and instrumentation for control' (Peter Peregrinus) Chap. 5

ROOTS, W. K. (1969): 'Fundamentals of temperature control' (Academic Press)

SCHMIDT, F. W., HENDERSON, R. E., and WOLGEMUTH, C. H. (1984): 'Introduction to thermal sciences' (John Wiley)

SHINSKEY, F. G. (1978): 'Energy conservation through control' (Academic Press)

STEPHANOPOULOS, G. (1984): 'Chemical process control' (Prentice–Hall International)

UNBEHAUEN, H., SCHMID, Chr., and BÖTTIGER, F. (1976): 'Comparison and application of DDC algorithms for a heat exchanger', *Automatica*, **12**, pp. 393–402

URONEN, P., and YLINIEMI, L. (1977): 'Experimental comparison and application of different DDC algorithms' *in* 'Digital computer applications to process control' Proceedings of the IFAC Conference, pp. 457–464

WILLIAMS, A. H., and WADDINGTON, J. (1983): 'New automatic control strategies for CEGB boilers', *CEGB Research*, pp. 16–24

ZEMANSKY, M.W., and DITTMAN, R. M. (1981): 'Heat and thermodynamics' (McGraw–Hill)

Appendix

Review of some basic definitions and some thermodynamic fundamentals

A.1 Introduction

Anyone proposing to work on the measurement and control of thermal processes will probably have some knowledge of the theoretical foundations of thermal phenomena. The rapid review that follows is intended to set out the salient features of theory that govern system behaviour. There are many excellent texts (a few are quoted in the list of references, Section A.19) covering thermodynamic fundamentals at different levels and from different viewpoints.

A.2 Laws of thermodynamics

Zeroth law: If two systems are separately in thermal equilibrium with a third system then they are in thermal equilibrium with each other.

First law: Let a system in state a have total energy U_a and let the system be brought to a new state b with corresponding total energy U_b, then the first law states that:

$$U_b - U_a = K = W + Q \tag{A1}$$

where K is a constant depending only on a and b, W, Q are the work done and the heat energy added to the system, respectively, to bring it from state a to state b. If $Q = 0$, no heat enters or leaves the system and the change in state is adiabatic; If $W = 0$, the change of state is defined to be isochoric.

Second law: It is impossible for any process to have as its sole result the transfer of heat from a cooler to a hotter body. (Work may be dissipated completely into heat but heat may not be converted entirely into work.) The law may also be stated in the form due to Caratheodory which does not require any process to be envisaged: In the neighbourhood of every equilibrium state of a system, there exist states that cannot be reached by means of an adiabatic change.

A.2.1 First law: Illustration

Suppose that a system is supplied with a small quantity of heat dQ that is partly used to increase the internal energy of the system and partly used to do work on the external environment. We write

$$dQ = dU + dW \tag{A2}$$

Let the system now, by some means, be brought back to its initial condition. It is then said to have completed a *closed cycle*. The internal energy of the system is the same as at the start, since the conditions are identical. Hence we write

$$dQ = dW \tag{A3}$$

i.e. every time a system goes through a closed cycle, the heat energy expended equals the quantity of work obtained. This statement may be considered to be a form of the first law of thermodynamics.

A.2.2 Second law of thermodynamics: In terms of Carnot cycles

Fig. A1 shows a Carnot cycle in which a perfect gas performs a closed cycle of isothermal expansion, adiabatic expansion, isothermal compression and

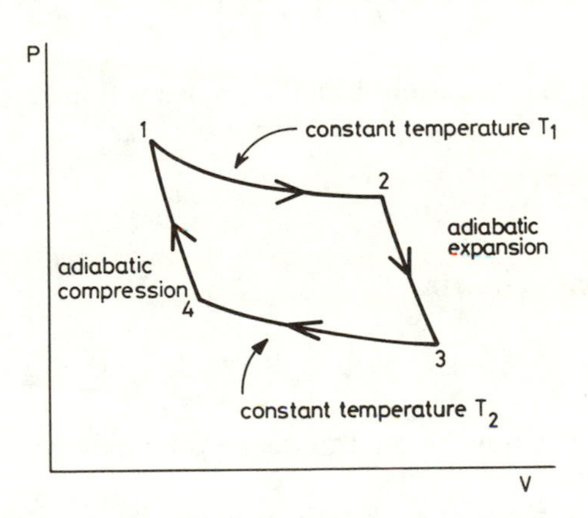

Fig. A1 *Carnot cycle*

adiabatic compression. The work involved in the two adiabatic changes is given by

$$W = MC_v(T_1 - T_2), \quad W = MC_v(T_2 - T_1) \quad \text{respectively.}$$

Hence no net work is done by the adiabatic changes over the complete cycle.

The work due to the two isothermal changes is given by

$$W = RT_1 \ln \frac{V_2}{V_1} - RT_2 \ln \frac{V_3}{V_4} \tag{A4}$$

However,

$$\frac{T_2}{T_1} = \left(\frac{V_2}{V_3}\right)\frac{C_p}{C_v} - 1 \quad \text{and} \quad \frac{T_1}{T_2} = \left(\frac{V_4}{V_1}\right)\frac{C_p}{C_v} - 1$$

hence

$$\frac{V_3}{V_4} = \frac{V_2}{V_1},$$

from which

$$W = R(T_1 - T_2) \ln \frac{V_2}{V_1} \tag{A5}$$

This is the net work produced in the cycle. Now the heat input to produce the work is equal by definition to the work done in the first isothermal expansion (a consequence of eqn. A4). Hence the heat input is $RT_1 \ln V_2/V_1$, and hence the efficiency of the cycle is

$$\frac{\text{work obtained}}{\text{energy input}} = \frac{R(T_1 - T_2) \ln \left(\frac{V_2}{V_1}\right)}{RT_1 \ln \left(\frac{V_2}{V_1}\right)} = \frac{T_1 - T_2}{T_1} \tag{A6}$$

But since in isothermal expansions heat is completely transformed into work, we can also express the efficiency of the Carnot cycle as

$$\eta = \frac{Q_1 - Q_2}{Q_1} \tag{A7}$$

and by comparison we obtain

$$\frac{T_1}{T_2} = \frac{Q_1}{Q_2} \tag{A8}$$

The Carnot cycle may now be used to define thermodynamic temperature.

A.2.3 Importance of the laws of thermodynamics

The pattern of evolution of physics has followed a general trend in which many individual laws and postulates have been replaced by a smaller number of more fundamental, more all-embracing laws and postulates. In this process of evolution, which is clearly motivated by the desire to account for all observable phenomena within a single theoretical scheme, a number of theorems were either overthrown or shown to be of only restricted validity. For instance, the developments of quantum theory and relativity theory both necessitated drastic revisions in the interlocking theorems of theoretical physics.

Through all the changes, the laws of thermodynamics emerged to form a central and fundamental part of the increasingly unified framework of theoretical physics. The first law has had to be generalised to take account of atomic energy.

A.2.4 Meaning of the laws of thermodynamics

The first law of classical thermodynamics is a statement that the total energy in a closed system remains constant regardless of changes that take place within the system. However, in the production of atomic energy, mass is actually converted into energy according to the relation $e = mc^2$, where m is the rest mass of the material converted. The generalised first law states that the sum of the energy and thc energy equivalent of the rest mass in a closed system remains constant regardless of changes that take placc within the system.

The second law of thermodynamics is essentially an inequality. The law indicates that there are irreversible processes and it defines the direction in which such processes much monotonically evolve. The second law is a probabilistic law in the sense that it requires that a system must evolve in time from any current state towards a most probable state. The second law does not treat individual microscopic elements deterministically, but rather treats macroscopic sets of elements statistically and concerns itself with most probable directions of evolution. The second law has particular philosophical implications, for it is the only law that distinguishes 'past' from 'future' and which thus allows the positive direction of the time axis to be fixed. The second law is usually stated as an inequality involving the quantity 'entropy'. Entropy will be defined later, but, for the moment, we state that every closed system has associated with it a scalar-valued non-negative quantity – its entropy. Entropy is not an invariant – it can change according to events within the system. The second law states that every closed system will evolve in such a way that its entropy increases. Recall that the second law is a probabilistic law, so that what is being said is that the preferred state of a closed system is that state for which the entropy is maximum.

A.3 Heat

Heat is energy in transit due solely to temperature differences. Only the rate $\dot{Q}$ oif heat flow can actually be determined at an instant of time, and then the quantity Q of heat may be calculated by integration between definite time limits, say 0 and t_1 as below:

$$\text{i.e.} \quad Q = \int_0^{t_1} \dot{Q}\, dt \qquad \text{(A9)}$$

As a result of heat exchange, the internal energy U in a body can change according to the law

$$U_T = U_0 + \int_0^{t_1} \dot{Q}\, dt \qquad \text{(A10)}$$

where the integral on the right-hand side is path dependent, i.e. $\dot{Q}\, dt$ is an inexact differential.

Where heat is exchanged between two bodies without losses to the surroundings then the heat lost by body $A(=Q_A)$ is equal to the heat gained by body $B(=Q_B)$

$$\text{i.e.} \quad Q_A + Q_B = 0 \qquad \text{(A11)}$$

A.4 Enthalpy

Consider a constant mass of gas at pressure P, volume V and internal energy U; we define the enthalpy of the gas by

$$H = U + PV \qquad \text{(A12)}$$

Latent heat is the change in enthalpy during a change of state.

A.5 Heat capacity

When heat is transferred to a system, the temperature of the system may or may not change. In those cases where an increment of heat dq causes a change $d\theta$ of temperature, we define the *heat capacity* of the system as

$$C = \frac{dq}{d\theta} \qquad \text{(A13)}$$

The heat capacity per unit mass of a substance is usually called its *specific heat*. The heat capacity per mole of substance is called the *molar specific heat*.

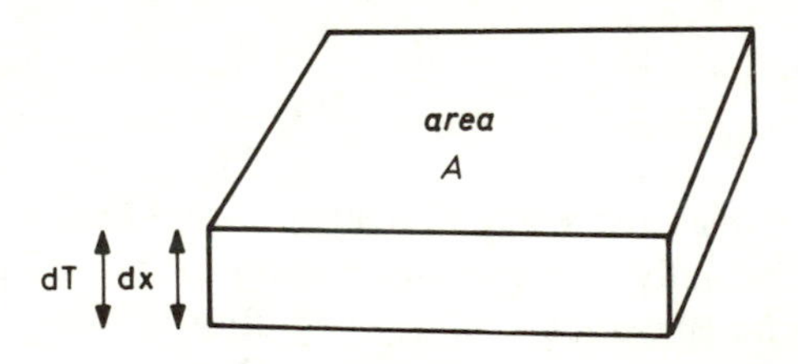

Fig. A2 *A slab of material across which a temperature gradient exists*

For any simple system, the heat capacity is a function of two variables, and for practical convenience we fix one or other of these and then regard heat capacity as a function of the single remaining variable. For instance, for gases, heat capacity is a function of both pressure and volume. The heat capacity at constant pressure is C_p and at constant volume is C_v.

A.6 Heat conduction

Fig. A2 shows a slab of material of thickness dx and cross-sectional area A across which a temperature difference dT is somehow maintained. The *temperature gradient* across the material is defined to be dT/dx. The rate of heat conduction through the slab is given by

$$\frac{dQ}{dt} = KA\frac{dT}{dx} \tag{A14}$$

where K is the *thermal conductivity* of the material in the slab; K is a nonlinear function of temperature whose form varies widely for different substances.

A.7 Convection

In heat transfer by convection, an intermediate moving fluid (liquid or gas) is involved as a transfer medium. Convection that takes place because of fluid movement caused by temperature-dependent changes of fluid density is called *natural convection*. When the convection currents are induced artificially, the phenomenon is called *forced convection*. Let a wall have an area A and let a temperature difference dT exist between the wall and an adjacent fluid; then the rate of heat transfer from the wall due solely to convection is given by

$$\frac{dQ}{dt} = hA\,dT \tag{A15}$$

where h is a difficult-to-determine coefficient, dependent on the physical state and condition of the wall and of the fluid.

A.8 Radiation: Stefan–Boltzmann's law

The Stefan–Boltzmann law quantifies the heat transfer from a body of surface area A and emissivity α at a temperature T to totally surrounding walls at temperature T_w as

$$\dot{Q} = A\alpha\varrho(T^4 - T_w^4) \tag{A16}$$

where ϱ is the Stefan–Boltzmann constant. It has the value

$$\varrho = 5{\cdot}67 \times 10^{-8}\ \mathrm{W/m^2\,K^4}$$

A.9 Internal energy

For simple systems, not consisting of mixed physical phenomena, the internal energy U is always a function of any two of the variables X_1, X_2, X_2 where X_1,

X_2 are application-dependent variables but X_3 is always temperature. For instance, where the system is a gas, we have X_1 = pressure, X_2 = volume, X_3 = temperature, or where the system is an electric cell we have X_1 = EMF, X_2 = charge, X_3 = temperature.

The first law, in differential form is

$$dU = dQ + dW$$

and in terms of the X co-ordinates just discussed,

$$dU = dQ + X\,dX_2 \qquad \text{(For a gas } dU = dQ + P\,dV\text{)} \qquad \text{(A17)}$$

Although dQ from eqn. A17 is an inexact differential, it can be integrated after multiplying by an integrating factor *that always exists for every function of two variables.*

For composite systems, dQ in general is an inexact differential that is a function of three or more independent variables. Now, whereas an integrating factor always exists for any differential function of two variables, no integrating factor exists in general when dQ is a function of three or more variables. However, it is a consequence of the second law of thermodynamics that every dQ does have an integrating factor. Further, the integrating factor is the same function, of temperature only, for all systems. This fact enables a link to be established between fundamental processes and an absolute temperature scale.

A.10 Condition for a function to be an exact differential

Let $f(x, y)$ be a differentiable function of both x and y, then

$$df(x, y) = \frac{\partial f}{\partial x}dx + \frac{\partial f}{\partial y}dy \qquad \text{(A18)}$$

and if we have the condition

$$\frac{\partial}{\partial y}\left(\frac{\partial f}{\partial x}\right) = \frac{\partial^2 f}{\partial y \partial x} = \frac{\partial^2 f}{\partial x \partial y} = \frac{\partial}{\partial x}\left(\frac{\partial f}{\partial y}\right) \qquad \text{(A19)}$$

then df is an *exact differential* of some function f.

When an exact differential is integrated between specified limits (represented in the above example by specified points [(x_1, y_1), (x_2, y_2) in the x, y plane)] then the result depends only on the limits [in the example, the result is $f(x_2, y_2) - f(x_1, x_2)$, which is seen to be *path independent*].

The integrals of exact differentials are often found to be important physical variables. For instance, the work done in moving a mass in a gravitational field is an exact differential, and hence there exists a function (the potential energy) the difference of whose values at the defined end points of the movement is equal to the work performed. Similarly the work done in moving an electric charge in an electric field is path independent and implies that the electrical potential function must exist. Exact differentials with their implications of path independence and the existence of potential-like functions will be encountered again in the definition of entropy.

Suppose that a system surrounded by an adiabatic boundary can nevertheless have work performed on it through the adiabatic boundary to take it from an initial state to a final state, then the work done depends only on the initial and final states of some function U of the system and is path independent, i.e. work done in moving from state 1 to state 2 is

$$W = U_2 - U_1 \tag{A20}$$

U is the *internal energy* in the system and the statement can be considered as one implication of the *first law of thermodynamics*.

The first law is a statement guaranteeing the conservation of energy and the existence of an internal energy function. The first law also shows the equivalence of heat and work. For a general system surrounded by a diathermal wall and subject both to work and heat transfer from the environment, the first law states that

$$U_2 - U_1 = W + Q \tag{A21}$$

i.e. the change in internal energy is the sum of work done and heat transferred. The history of the development of the understanding of the quantitative relation expressed in eq. A21 is fascinating: it involved such experiments as drilling the bores of cannons with blunt tools to determine the heat produced.

A.11 Entropy

It can be seen that, in terms of conversion to useful work, an element of heat dQ is most valuable if it is available at a high temperature. This concept is quantified by the use of the *entropy*. Let a system take a quantity of heat dQ at temperature T from a source; then the entropy of the system is defined to increase by an amount $dS = dQ/T$, and in going from state a to state b, the entropy change will be

$$S_b - S_a = \int_a^b \frac{dQ}{T} \tag{A22}$$

Any system going through a closed reversible cycle suffers no change in entropy over a cycle since, from eqn. A8,

$$\frac{T_1}{T_2} = \frac{Q_1}{Q_2} \tag{A23}$$

Consider the closed cycle of Fig. A3. From what we have concluded, around the complete cycle,

$$\oint \frac{dQ}{T} = 0 \tag{A24}$$

However,

$$\oint \frac{dQ}{T} = \underset{\text{over segment 1}}{\int_A^B \frac{dQ}{T}} + \underset{\text{over segment 2}}{\int_B^A \frac{dQ}{T}}$$

However,

$$\underset{\text{over segment 2}}{\int_B^A \frac{dQ}{T}} = -\underset{\text{over segment 2}}{\int_A^B \frac{dQ}{T}}$$

and hence

$$\underset{\text{over segment 1}}{\int_A^B \frac{dQ}{T}} = \underset{\text{over segment 2}}{\int_A^B \frac{dQ}{T}} \tag{A25}$$

i.e. the entropy increase from A to B does not depend on the path provided that the process is reversible. The entropy at a point is a function only of the point, i.e. *entropy is a state variable.*

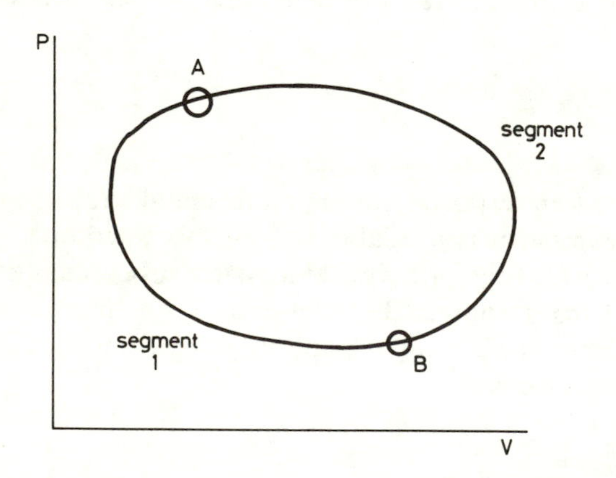

Fig. A3 *A closed cycle in the P/V plane*

The entropy of a perfect gas can be obtained as follows: From Section A.2 it follows that

$$dQ = MC_v dT + RT\,dV/V \tag{A26}$$

Therefore

$$dS = MC_v dT + \frac{R\,dV}{V} \tag{A27}$$

Integrating from (T_0, V_0) to (T, V) leads to

$$S = MC_v \ln\frac{T}{T_0} + R\ln\frac{V}{V_0}$$

which simplifies to

$$S = MC_v \ln T + R\ln V + \text{a constant} \tag{A28}$$

and this is the entropy of a perfect gas that is in the state (T, V).

If we consider an isolated reversible system it is clear that its entropy must be constant even though there may be changes taking place in the internal distribution of entropy within the system.

If we now consider an isolated irrversible system we find that the entropy can only increase towards a final maximum value.

A.11.1 Simple illustrative example

Let 1 kg of water at 100°C be mixed with 1 kg of water at 0°C. What is the change in entropy?

$$dS = 1000\left(\ln\frac{323}{273} + \ln\frac{323}{373}\right)$$

$$= 168 - 144 = 24\,\text{cal}\,\text{K}^{-1}$$

A.12 Reversible and irreversible processes

As we have seen earlier, macroscopic processes may be divided into *reversible* and *irreversible processes*. If $f(t)$ is a reversible process, then the process $f(-t)$ is also possible. If $f(t)$ is an irreversible process then the process $f(-t)$ is impossible. A planet revolving in the solar system or a light ray being refracted are examples of reversible systems.

A.12.1 Statistical irreversibility

The most probable state in which a gas can exist is the equilibrium state. This means that, after a long time has elapsed, the probability of finding the gas in a state other than its most probable state is virtually zero. (The very large size of Avogadros's number makes the probability curve very sharp.) In the equilibrium state, the entropy will be maximum – its value can be determined from Maxwell's equation describing the velocity distribution of the molecules of the gas.

The second law of thermodynamics states that a macroscopic system progresses monotonically from the less probable to the more probable condition.

The most probable state is the equilibrium state. Notice that there is microscopic reversibility but macroscopic irreversibility. Time t_2 is after time t_1 if for the overwhelming majority of cases of isolated systems we find that $S(t_2) > S(t_1)$ (Brownian motion goes on even at equilibrium, but no work can be extracted from it).

A.13 Kinetic theory of gases

The theory is, in outline, that a gas is made up of rapidly moving molecules, that the heat contained in the gas is exactly the kinetic energy of the molecules and that the pressure on the container walls is produced by the collision and recoil of molecules with those walls.

Using this outline theory together with fairly simplistic assumptions about the distribution of molecules and their velocities, it is possible to predict numerical values for physical variables, using momentum and energy arguments. For instance, from first principles, we obtain for a perfect gas that $C_p/C_v = 5/3$, which agrees well with the experimentally obtained value of $C_p/C_v = 1{\cdot}67$ for monatomic gases. The distribution of velocities of the molecules in a gas is not uniform: rather (for a rarified gas) it is described by Maxwell's formula

$$F(V_x, V_y, V_z) = \frac{N}{V}\frac{m}{2\pi kT}e^{-mV^2/2kT} \tag{A29}$$

where

N = number of molecules
V = volume
T = temperature
m = mass of a molecule
k = Boltzmann's constant

Eqn. A29 allows all the properties of the gas, such as mean free path of the molecules, viscosity and thermal conductivity, to be derived.

A.14 *P/V* and *P/θ* diagrams

Fig. A4 shows a pressure/volume diagram for a typical pure substance, and Fig. A5 shows a pressure/temperature diagram for a typical pure substance.

A.15 Work and the *P/V* diagram

Consider a gas being taken from state A to state B along three alternative paths

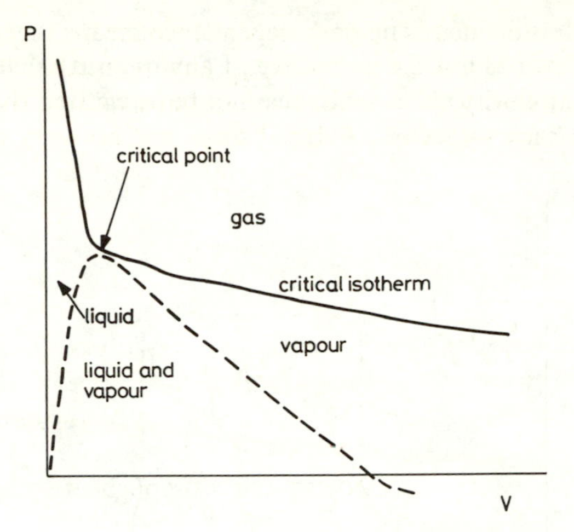

Fig. A4 *Isotherm diagram for a pure substance*

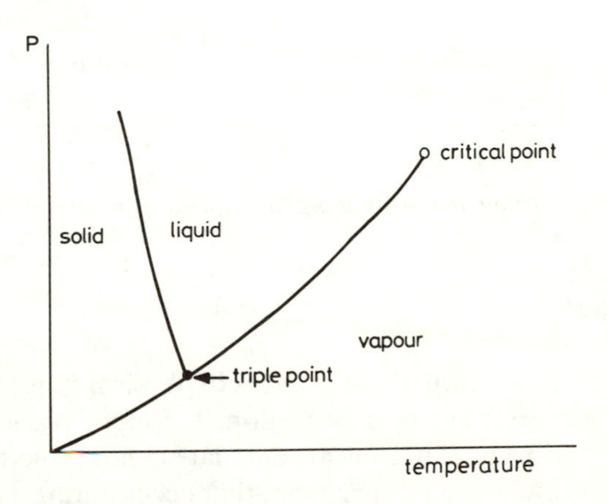

Fig. A5 *Typical pressure/temperature curve*

ACB, AOB, ADB in the P/V diagram, (Fig. A6). The work done on the gas is, in the three cases

$$P_2(V_2 - V_1), \quad \left(\frac{P_2 + P_1}{2}\right)\left(\frac{V_1 + V_2}{2}\right), \quad P(V_2 - V_1)$$

being the area under the respective paths. In other words, the work done is *path dependent*.

Mathematically the expression for the work done in going from A to B is given by

$$\int_{V_1}^{V_2} p(v)\,dv \tag{A30}$$

and the integration can only be performed when p is defined as a function of v, i.e. when the path is defined. The path dependence means that the infinitesimal quantity of work $p\,dv$ is not the derivative of any quantity that could be called the total work. A quantity like $p\,dv$ that cannot be integrated is called an *inexact differential* (refer back to Section A.10).

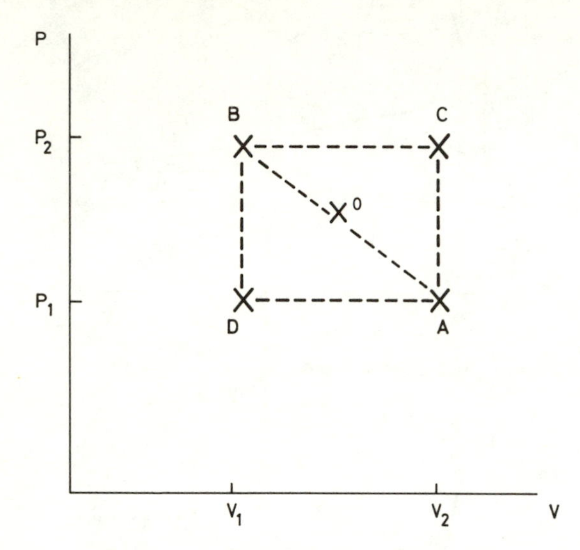

Fig. A6 *Diagram to illustrate that work is a path-dependent quantity*

A.16 Phase transitions

By 'phase transitions' we mean those changes of physical state such as take place during melting, vaporisation or sublimation. During a phase transition temperature remains constant, but, because of latent heat, there is a change of entropy. Fig. A7 shows the type of changes that occur during a phase transition.

Let ΔV be the total volume change that occurs in phase transition; then *Clapeyron's equation* states that

$$\frac{dP}{dT} = \frac{1}{T\Delta V} \tag{A31}$$

A.17 Maxwell's relations

We define the *Helmholtz function F* by

$$F = U - TS \tag{A32}$$

and the *Gibbs function G* by

$$G = H - TS \tag{A33}$$

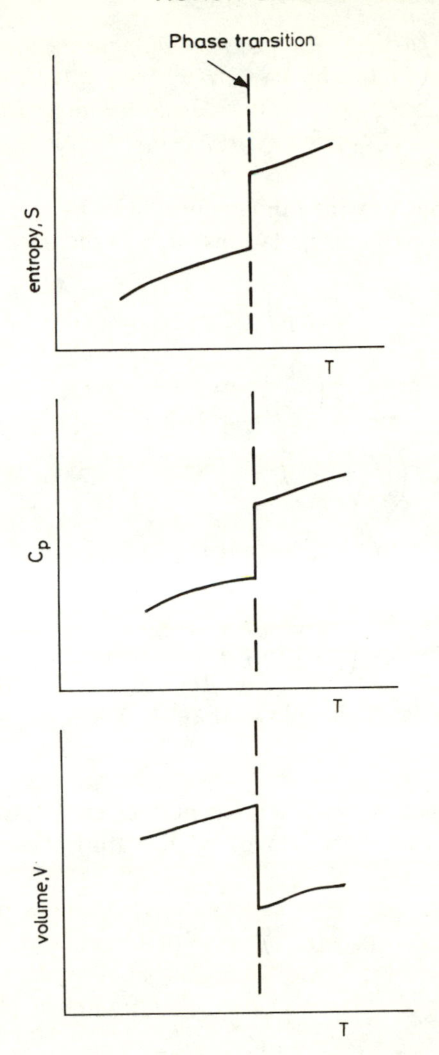

Fig. A7 *Typical changes in entropy S, heat capacity C_P and volume V during a phase transition*

and together with the internal energy U and the enthalpy $H = U + PV$ we can set up *Maxwell's relations*.

At an equilibrium state, the following relations hold amongst the four quantities defined above:

$$\left(\frac{\partial T}{\partial V}\right)_S = -\left(\frac{\partial P}{\partial S}\right)_V \tag{A34}$$

$$\left(\frac{\partial T}{\partial P}\right)_S = \left(\frac{\partial V}{\partial S}\right)_P \tag{A35}$$

$$\left(\frac{\partial S}{\partial V}\right)_T = \left(\frac{\partial P}{\partial T}\right)_V \quad \text{(A36)}$$

$$\left(\frac{\partial S}{\partial P}\right)_T = -\left(\frac{\partial V}{\partial T}\right)_P \quad \text{(A37)}$$

The Maxwell relations provide fundamental links between the thermodynamic quantities and allow particular relations such as the ones below to be derived:

$$TdS = C_v dT + T\left(\frac{\partial P}{\partial T}\right)_V dV \quad \text{(A38)}$$

$$TdS = C_p dT - T\left(\frac{\partial V}{\partial T}\right)_P dP \quad \text{(A39)}$$

$$\left(\frac{\partial U}{\partial V}\right)_T = T\left(\frac{\partial P}{\partial T}\right)_V - P \quad \text{(A40)}$$

$$\left(\frac{\partial U}{\partial P}\right)_T = -T\left(\frac{\partial V}{\partial T}\right)_P - P\left(\frac{\partial V}{\partial P}\right)_T \quad \text{(A41)}$$

$$C_p - C_v = -T\left(\frac{\partial V}{\partial T}\right)_P^2\left(\frac{\partial P}{\partial V}\right)_T \quad \text{(A42)}$$

And from this last relation it follows that

$$C_p \geqslant C_v \quad \text{(A43)}$$

This agrees with intuition: heating a gas at constant pressure involves work as the gas expands. No such work is involved in heating at constant volume. It also follows that

$$(C_p - C_v) \longrightarrow 0 \quad \text{as} \quad T \longrightarrow 0$$

and that

$$C_p = C_v \quad \text{whenever} \quad \frac{\partial V}{\partial T} = 0$$

A.18 Onsager's reciprocal relation

Suppose that a metal wire is subject, along its length, both to a temperature difference ΔT and an electrical potential difference ΔE, then a heat current I_H and an electric current I_C flow along the wire. Entropy is generated in the wire by both the temperature difference and the potential difference according to the relation

$$\frac{dS}{dt} = I_H \frac{\Delta T}{T} + I_C \frac{\Delta E}{T} \quad \text{(A44)}$$

and provided that ΔT and ΔE are relatively small, the following linear relations hold:

$$I_H = L_{11}\frac{\Delta T}{T} + L_{12}\frac{\Delta E}{T} \tag{A45}$$

$$I_C = L_{21}\frac{\Delta T}{T} + L_{22}\frac{\Delta E}{T} \tag{A46}$$

These are the *Onsager equations* and their L coefficients are associated with thermal conductivity, electrical conductivity and thermoelectricity. A final important relation amongst the coefficients in eqns. A45 and A46 is that $L_{12} = L_{21}$, which is known as the *Onsager reciprocal relation.*

A.19 References and further reading

ARPACI, V. S., and LARSEN, P. S. (1985): 'Convection heat transfer' (Prentice–Hall)

BENZINGER, T. H. (1964): 'Thermodynamics of life and growth' *in* HEALD, F, P. (Ed.): 'Adolescent nutrition and growth' (Appleton, Century, Croft, New York)

BENZINGER, T. H. (1971): 'Thermodynamics, chemical reactions and molecular biology', *Nature,* **229,** p. 100

BENZINGER, T. H. (1982): 'Temperature and thermodynamics of living matter' *in* SCHOOLEY, J. F. (Ed.): 'Temperature: its measurement and control in science and industry'. Vol. 5 (American Institute of Physics)

BUCHDAHL, H. A. (1975): 'Twenty lectures on thermodynamics' (Pergamon Press)

CALLEN, H. B. (1948): 'The application of Onsager's reciprocal relations to thermoelectric, thermomagnetic and galvanomagnetic effects', *Phys. Rev.* **73,** pp. 1349–1358

CANNON, J. R. (1984): 'The one-dimensional heat equation' (Cambridge University Press)

CHAPMAN, A. J. (1984): 'Heat transfer' (Macmillan)

GUILDNER, L. A. (1982): 'The measurement of thermodynamic temperature' *in* SCHOOLEY, J. F. (Ed.): 'Temperature: its measurement and control in science and industry'. Vol. 5 (American Institute of Physics)

GRIGULL, U., and SANDNER, H. (1984): 'Heat conduction' (Springer–Verlag)

HOLMAN, J. P. (1981): 'Heat transfer' (McGraw–Hill)

KAKAC, S. (1982): 'Heat transfer' (McGraw–Hill)

KYLE, B. G. (1984): 'Chemical and process thermodynamics' (Prentice–Hall)

LEWIS, R. W. (1981): 'Numerical methods in heat transfer' (John Wiley)

MARTIN, N. F. G., and ENGLAND, J. W. (1981): 'Mathematical theory of entropy' (Cambridge University Press)

MODELL, M., and REID, R.C. (1983): 'Thermodynamics and its applications' (Prentice–Hall)

RUELLE, D. (1978): 'Thermodynamic formalism (Cambridge University Press)

SMITH, E. B. (1982): 'Basic chemical thermodynamics' (Clarendon Press)

SOLOUKHIN, R. I., and AFGAN, N. (Ed.) (1984): 'Measurement techniques in heat and mass transfer' (Springer–Verlag)

SONNTAG, R. E., and VAN WYLEN, G. J. (1982): 'Introduction to thermodynamics: classical and statistical' (John Wiley)

STURTEVANT, J. M. (1977): 'Heat capacity and entropy changes in processes involving proteins' *Proc. Nat. Acad. Sci.,* **74,** p. 2236

ZEMANSKY, M. W., and DITTMAN, R. M. (1981): 'Heat and thermodynamics' (McGraw–Hill)

Index